SMART Integrated Circuit Design and Methodology

Editors

Thomas Noulis
Aristotle University of Thessaloniki, Greece

Costas Psychalinos
University of Patras, Greece

Alkis Hatzopoulos
Aristotle University of Thessaloniki, Greece

Tutorials in Circuits and Systems

For a list of other books in this series, visit www.riverpublishers.com

Series Editors

Manuel Delgado-Restituto
IEEE CASS President

Xinmiao Zhang
VP - Technical Activities, IEEE CASS

Kea-Tiong (Samuel) Tang
VP - Regional Activities and Membership, IEEE CASS

Published 2024 by River Publishers
Alsbjergvej 10
9260 Gistrup Denmark

www.riverpublishers.com

Distributed exclusively by Routledge
605 Third Avenue, New York, NY 10017, USA
4 Park Square, Milton Park, Abingdon, Oxon OX14 4RN

ISBN: 978-87-7022-833-6 (hardback)
ISBN: 978-10-0382-809-9 (online)
ISBN: 978-1-032-66558-0 (master ebook)

Library of Congress Cataloging-in-Publication Data: May 2024
Editors: Thomas Noulis, Costas Psychalinos and Alkis A. Hatzopoulos
Title: *SMART Integrated Circuit Design and Methodology*

Table of contents

Introduction

Systems-on-chip are available in every electronic product, and especially in emerging market segments such as 5G mobile communications, autonomous driving and fully electrified vehicles, and artificial intelligence. These complex product types require real-time processing at billions of operations per second. SoC complexity is rising, and circuit and system design groups are under extreme pressure to deliver efficient and competitive products, with optimized area to cost ratio and shorter design cycle times than ever before. The traditional design methodologies have reached their limits and innovative solutions are essential as to serve the emerging SoC design challenges.

Integrated Circuit and System design flow begins with the integration process where the selected devices or blocks are interconnected. The target is achieving the best power performance, fulfilling each application specs versus noise and speed and reaching the required area to cost ratio spec to have a viable product business case. Another critical factor is the design cycle time since this determines the design development cost and the time to market. In the standard design flow, each design task — power consumption, noise, speed, architecture, testing, and so on — is performed separately by specialized engineering teams. Close communications and continual information exchange is required among the teams. Each new design iteration restarts the information exchange process resulting to a high design cycle times and therefore increasing dramatically the project's final cost, time to completion and business case validity.

Innovative and smart circuit and system design methodologies are required as to move from the standard traditional SoC methodology to an up-to-date methodology that can face the current market needs and challenges. This is essential as to accomplish higher number of design tasks, without stretching engineering resources and exploding development costs. Design methodology and related innovations should focus on pushing the limits and a scalable and boundary less way to accelerate the design process in the most efficient, convergent, and cost-effective manner should be defined. Artificial intelligence and machine learning should be adequately integrated in the design methodology resulting in faster convergence to optimized design flows, reduced margins, and thus a greater opportunity to realize aggressive design targets within shorter schedules.

In the framework of the Circuit and System Society (CASS) Outreach Initiative 2022 call, the SMART Integrated Circuits design methodology – named SMARTIC - Seasonal School was performed in November 2022, in Thessaloniki (Greece) and in KEDEA of Aristotle University of Thessaloniki. In this educational 3-day seasonal school, basic and advanced issues on SoC design methodology targeting applications from power management, filtering, digital processing to radiation detection interfaces and printed electronics were addressed. Several advanced flows and methodologies for the design and implementation of SoCs were covered, such as utilization of machine learning and artificial intelligence, gm/Id based analog design and approximation and acceleration.

In Chapter 1 and Chapter 2, core analog circuits of any System of Chip, such as high-performance rectifiers and filters, are addressed in detail, together with their respective design methodology. Analog ICs experts – Prof. Shahram Minaei and Prof. Costas Psychalinos - present advanced topologies and address all the related trade-offs in the implementation of such architectures.

In Chapter 3 and Chapter 4, way modern and advanced methodologies towards design cycle speed up are addressed. A renowned expert on design, optimization, and mathematical modeling of analog, mixed-signal, and RF integrated circuits -Prof. Paul Sotiriadis – addresses circuits and systems for Machine Learning and Artificial Intelligence applications. In addition, Prof. Zervakis, addresses approximation- acceleration and mainly Machine Learning Classification On the Printed Circuits landscape.

In Chapter 5, Prof. Hesham Omran, expert in sensing microsystems, mixed-signal integrated circuit design, and FPGA based systems, presents an advanced analog design methodology based on gm/Id and lock up tables. A powerful flow for enabling fast time to market analog circuit design focusing on baseband circuits.

In Chapter 6 and Chapter 7, the focus is on more exotic methodologies and applications. Dr. Pedro Filipe Leite Correia De Toledo depicts Digital-Based Analog Processing in Nanoscale CMOS ICs. This is a design methodology, way modern and effective on achieving high speed IC design. Dr. Konstantinos Moustakas addresses the design and development of depleted monolithic active pixel sensors for high-radiation applications, together with all the respective challenges of this application.

Finally, on the verification of system on chip and moving on the digital domain, a chapter oriented on the practical topic of digital verification is presented. Mrs Olivera Stojanovic', product owner of the Cogita tool by VTOOL Ltd, which is oriented on debugging and assisting verification engineers, presents the topic of digital verification, the related challenges and explains why the skills of digital verifications engineers are current on high demand.

In conclusion, this eBook is written by a mixture of industrial experts and key academic professors and researchers. The intended audience is not only students but also engineers with system-on-chip and semiconductors background working in the semiconductor industry. This content can also be used as material in any graduate course curriculum related to Integrated circuit and system design. This eBook is extremely useful material for anyone involved in the implementation of high-performance system-on chip. We do believe you will enjoy reading this Ebook, the same way we have enjoyed preparing the SMARTIC Seasonal School 2022, executing it and attending all these invited talks.

The Editors,
Thomas Noulis, Costas Psychalinos and Alkis Hatzopoulos
January 19th, 2022

CHAPTER 01

On the Realization of High Accuracy Rectifiers Based on Modern Active Elements

Prof. Shahram Minaei

Dogus University, Department of Electrical
and Electronics Engineering, Istanbul, Turkey

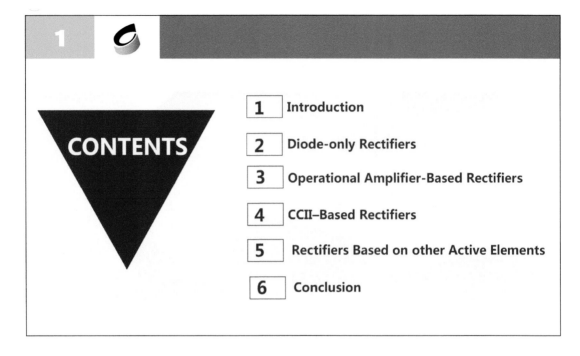

CONTENTS

1 Introduction

2 Diode-only Rectifiers

3 Operational Amplifier-Based Rectifiers

4 CCII–Based Rectifiers

5 Rectifiers Based on other Active Elements

6 Conclusion

SEC I: INTRODUCTION

- Threshold voltages of the diodes limit the applicable input signal amplitude in conventional DIODE-only rectifiers.

- On the other hand, for low-level signals, rectifiers with high precision are needed in instrumentation and measurement applications such as

  o AC voltmeter,
  o AC ammeter,
  o AC wattmeter,
  o function generators,
  o various nonlinear analog signal-processors,
  o averaging circuits,
  o peak value detectors,
  o …….etc.

Problems of the conventional diode-only rectifiers are discussed.
   Applications of high precision rectifiers are mentioned.

3

- So, there is a need to develop rectifiers with high accuracy for handling of signals with low amplitude.

- Many active elements such as,

  operational amplifier (OA),
  second-generation current conveyor (CCII),
  Second-generation voltage conveyors (VCII),
  current feedback operational amplifier (CFOA),
  differential voltage current conveyors (DVCC),
  operational transconductance amplifier (OTA),
  current differencing transconductance amplifier (CDTA)
       .....etc.,

  have been used to realize high precision rectifiers.

N eed of high accuracy rectifiers are emphasized.
 Different active elements that can be used for realization of high precision rectifiers are discussed.

4

- In this lecture first we start with threshold voltage (forward voltage drop, V_D) problem of diodes in conventional rectifiers.

- Then, different realizations of high accuracy rectifiers based on above mentioned active elements will be given.

R est of the slides are introduced.
 The slides continue with the discussing about the Forward voltage drop, V_D, of the diodes. Then high accuracy rectifiers will be given.

SEC II: DIODE-ONLY RECTIFIERS

5

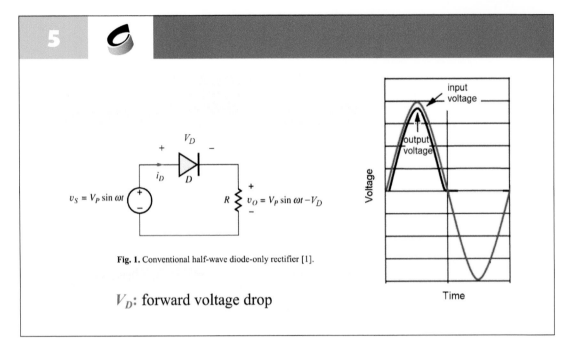

Fig. 1. Conventional half-wave diode-only rectifier [1].

V_D: forward voltage drop

A standard conventional half-wave diode-only rectifier is given.
Voltage difference between the input and output signals are discussed.

6

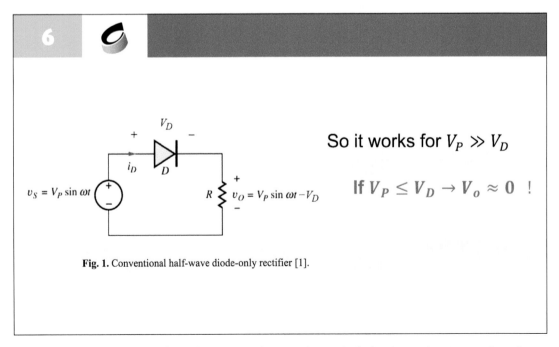

So it works for $V_P \gg V_D$

If $V_P \leq V_D \rightarrow V_o \approx 0$!

Fig. 1. Conventional half-wave diode-only rectifier [1].

By writing a KVL around the loop, the output voltage is obtained which is lower than input voltage by V_D.
If the value of V_D is too high, the output is zero.

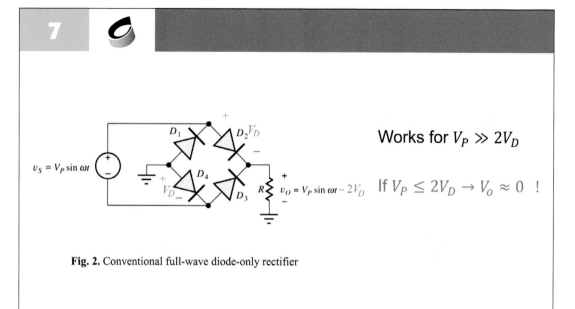

7

$v_S = V_P \sin \omega t$

D_1 D_2 V_D

D_4

V_D

D_3

R $v_O = V_P \sin \omega t - 2V_D$

Works for $V_P \gg 2V_D$

If $V_P \le 2V_D \rightarrow V_O \approx 0$!

Fig. 2. Conventional full-wave diode-only rectifier

In this slide a full wave rectifier (bridge rectifier) is given.
The forward voltage drop of diodes is totally equal to $2V_D$ in each cycle.

SEC III: OPERATIONAL AMPLIFIER-BASED RECTIFIERS

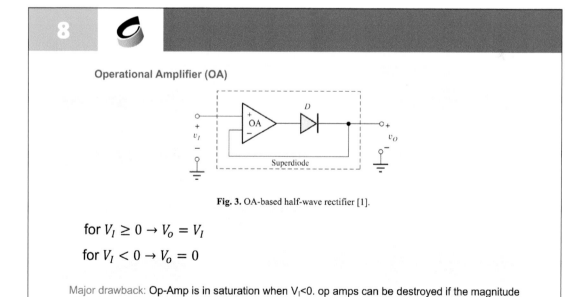

8

Operational Amplifier (OA)

D

OA

v_I

Superdiode

v_O

Fig. 3. OA-based half-wave rectifier [1].

for $V_I \ge 0 \rightarrow V_O = V_I$

for $V_I < 0 \rightarrow V_O = 0$

Major drawback: Op-Amp is in saturation when $V_I<0$. op amps can be destroyed if the magnitude of the input voltage is larger than a few volts.

Op-amp based high precision half-wave rectifier is given.
Major draw back of the circuit is discussed.

9

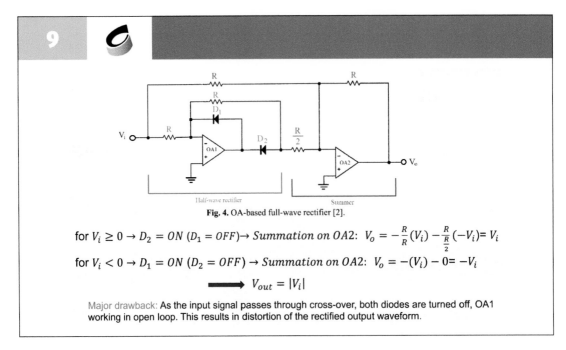

Fig. 4. OA-based full-wave rectifier [2].

for $V_i \geq 0 \rightarrow D_2 = ON \ (D_1 = OFF) \rightarrow Summation \ on \ OA2$: $V_o = -\frac{R}{R}(V_i) - \frac{R}{\frac{R}{2}}(-V_i) = V_i$

for $V_i < 0 \rightarrow D_1 = ON \ (D_2 = OFF) \rightarrow Summation \ on \ OA2$: $V_o = -(V_i) - 0 = -V_i$

$$\longrightarrow V_{out} = |V_i|$$

Major drawback: As the input signal passes through cross-over, both diodes are turned off, OA1 working in open loop. This results in distortion of the rectified output waveform.

A full-wave rectifier including op-amps and diodes is given.
  The circuit is composed of a half-wave rectifier and a summer circuit.
Major draw back of the circuit is discussed.

SEC IV: CCII–BASED RECTIFIERS

10

Another solution: **Pumping the diode with a current source**

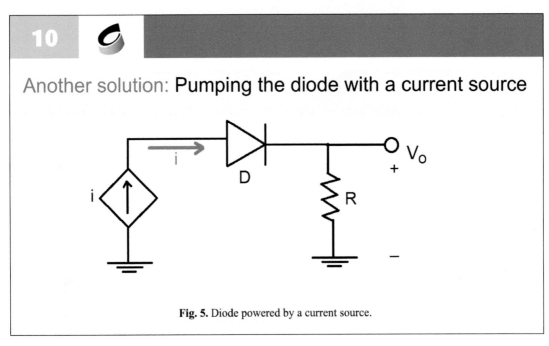

Fig. 5. Diode powered by a current source.

R ectification can be performed by a current source.
  Diode is pumped by a current source.

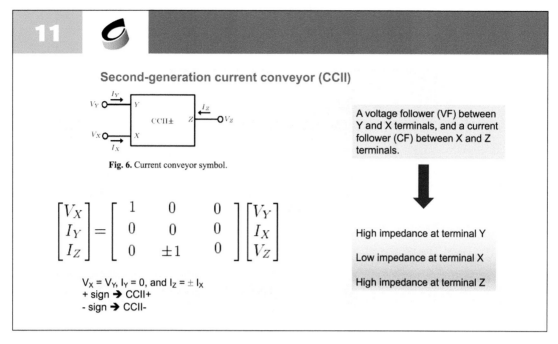

Second-generation current conveyor (CCII)

Fig. 6. Current conveyor symbol.

$$\begin{bmatrix} V_X \\ I_Y \\ I_Z \end{bmatrix} = \begin{bmatrix} 1 & 0 & 0 \\ 0 & 0 & 0 \\ 0 & \pm 1 & 0 \end{bmatrix} \begin{bmatrix} V_Y \\ I_X \\ V_Z \end{bmatrix}$$

$V_X = V_Y$, $I_Y = 0$, and $I_Z = \pm I_X$
+ sign → CCII+
- sign → CCII-

A voltage follower (VF) between Y and X terminals, and a current follower (CF) between X and Z terminals.

High impedance at terminal Y

Low impedance at terminal X

High impedance at terminal Z

Second-generation current conveyor is explained.
Its voltage-current relationship and impedances at different terminals are discussed.

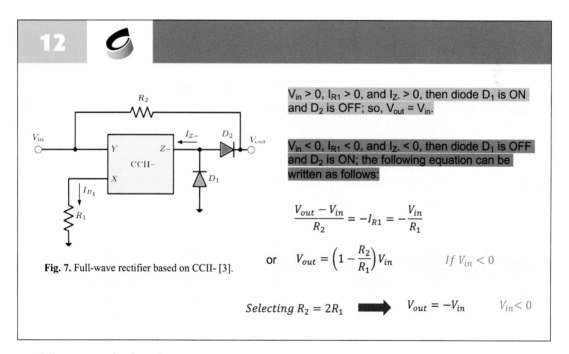

Fig. 7. Full-wave rectifier based on CCII- [3].

$V_{in} > 0$, $I_{R1} > 0$, and $I_{Z-} > 0$, then diode D_1 is ON and D_2 is OFF; so, $V_{out} = V_{in}$.

$V_{in} < 0$, $I_{R1} < 0$, and $I_{Z-} < 0$, then diode D_1 is OFF and D_2 is ON; the following equation can be written as follows:

$$\frac{V_{out} - V_{in}}{R_2} = -I_{R1} = -\frac{V_{in}}{R_1}$$

or $\quad V_{out} = \left(1 - \frac{R_2}{R_1}\right) V_{in} \qquad If\ V_{in} < 0$

Selecting $R_2 = 2R_1$ ➡ $V_{out} = -V_{in} \qquad V_{in} < 0$

A full-wave rectifier based on current conveyor is shown as an example.
The given circuit is analyzed in detail.

13

In other words:

$$V_{out} = |V_{in}|$$

Major drawback: **The proposed circuit has no high input and low output impedances, so buffer stage may be required at its input or output.**

The equation of the rectifier is given.
    Main drawback of the circuit is explained.

SEC V: RECTIFIERS BASED ON OTHER ACTIVE ELEMENTS

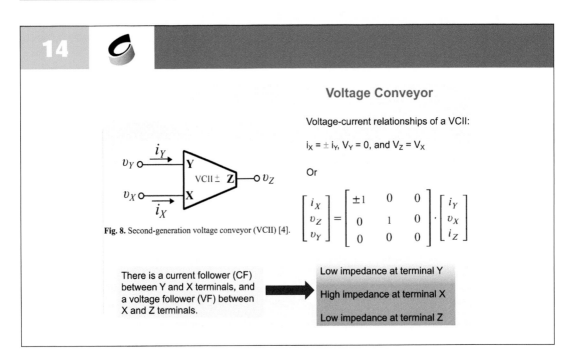

14

Voltage Conveyor

Voltage-current relationships of a VCII:

$i_X = \pm i_Y$, $V_Y = 0$, and $V_Z = V_X$

Or

$$\begin{bmatrix} i_X \\ v_Z \\ v_Y \end{bmatrix} = \begin{bmatrix} \pm 1 & 0 & 0 \\ 0 & 1 & 0 \\ 0 & 0 & 0 \end{bmatrix} \cdot \begin{bmatrix} i_Y \\ v_X \\ i_Z \end{bmatrix}$$

Fig. 8. Second-generation voltage conveyor (VCII) [4].

There is a current follower (CF) between Y and X terminals, and a voltage follower (VF) between X and Z terminals.

Low impedance at terminal Y

High impedance at terminal X

Low impedance at terminal Z

Second-generation voltage conveyor is described.
    Current-voltage relationships between terminals of the voltage conveyor are discussed.

15

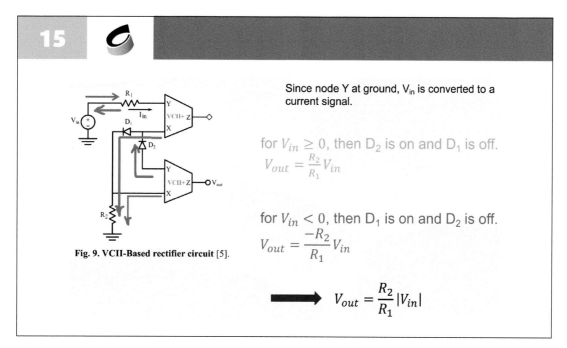

Since node Y at ground, V_{in} is converted to a current signal.

for $V_{in} \geq 0$, then D_2 is on and D_1 is off.
$$V_{out} = \frac{R_2}{R_1} V_{in}$$

for $V_{in} < 0$, then D_1 is on and D_2 is off.
$$V_{out} = \frac{-R_2}{R_1} V_{in}$$

Fig. 9. VCII-Based rectifier circuit [5].

$$V_{out} = \frac{R_2}{R_1} |V_{in}|$$

A full-wave rectifier based on voltage conveyor is given as an example. The circuit is described, and its output voltage is obtained. It is shown that the circuit has variable gain.

16

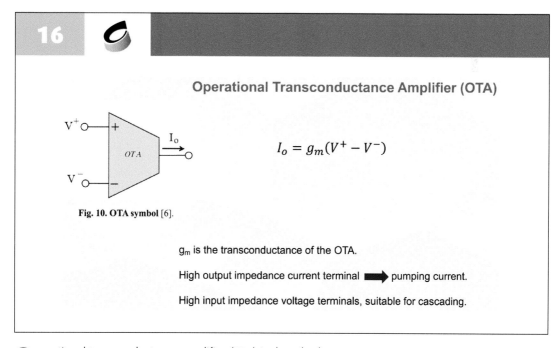

Operational Transconductance Amplifier (OTA)

$$I_o = g_m(V^+ - V^-)$$

Fig. 10. OTA symbol [6].

g_m is the transconductance of the OTA.

High output impedance current terminal ➡ pumping current.

High input impedance voltage terminals, suitable for cascading.

O perational transconductance amplifier (OTA) is described. Impedances of different terminals of the OTA are clarified, and output current equation is given.

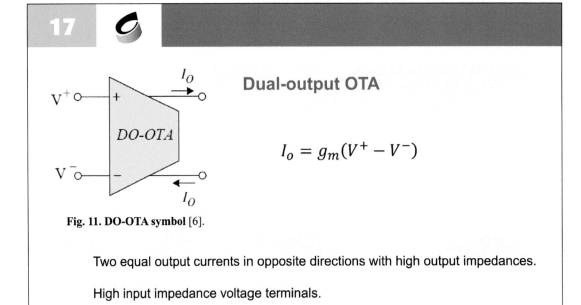

17

Dual-output OTA

$$I_o = g_m(V^+ - V^-)$$

Fig. 11. DO-OTA symbol [6].

Two equal output currents in opposite directions with high output impedances.

High input impedance voltage terminals.

The electrical symbol of the Dual-output operational transconductance amplifier is given. Directions of its output currents are described.

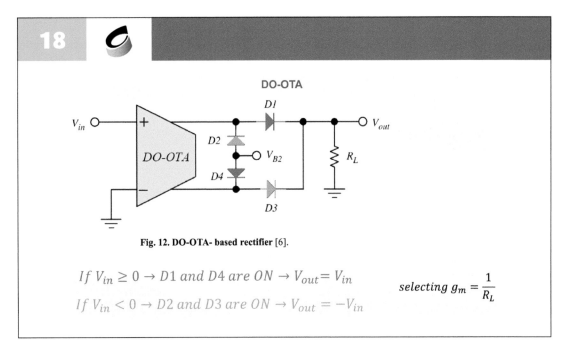

18

DO-OTA

Fig. 12. DO-OTA- based rectifier [6].

If $V_{in} \geq 0 \rightarrow D1$ and $D4$ are ON $\rightarrow V_{out} = V_{in}$

If $V_{in} < 0 \rightarrow D2$ and $D3$ are ON $\rightarrow V_{out} = -V_{in}$

selecting $g_m = \dfrac{1}{R_L}$

A full-wave rectifier based on dual output operational transconductance amplifier is shown. The operation of the circuit is explained.

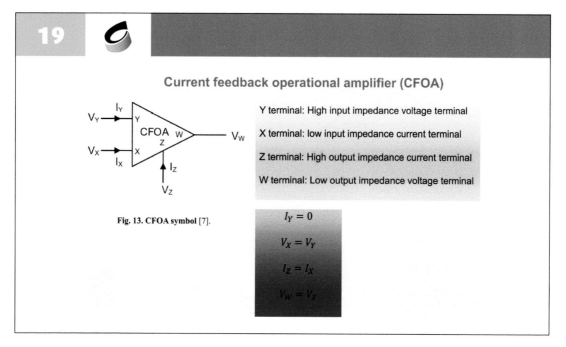

19

Current feedback operational amplifier (CFOA)

Y terminal: High input impedance voltage terminal

X terminal: low input impedance current terminal

Z terminal: High output impedance current terminal

W terminal: Low output impedance voltage terminal

Fig. 13. CFOA symbol [7].

$$I_Y = 0$$

$$V_X = V_Y$$

$$I_Z = I_X$$

$$V_W = V_Z$$

Current feedback operational amplifier (CFOA) as an active element is described.
Voltage-current relationships among the terminals of the CFOA are discussed.

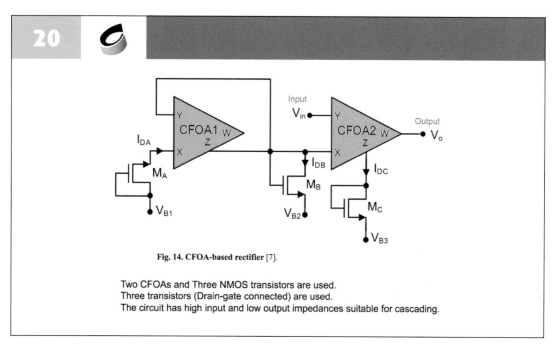

20

Fig. 14. CFOA-based rectifier [7].

Two CFOAs and Three NMOS transistors are used.
Three transistors (Drain-gate connected) are used.
The circuit has high input and low output impedances suitable for cascading.

A full-wave rectifier based on two CFOAs and three NMOS transistors is explained.
The input and output impedances of the circuit is clarified.

21

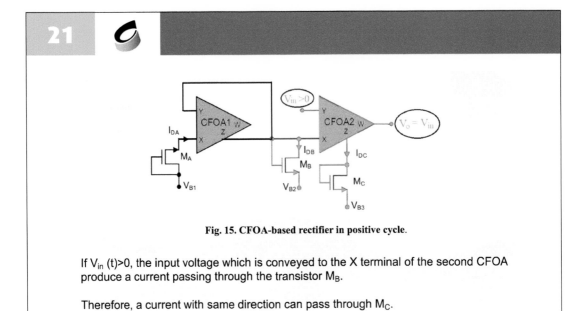

Fig. 15. CFOA-based rectifier in positive cycle.

If V_{in} (t)>0, the input voltage which is conveyed to the X terminal of the second CFOA produce a current passing through the transistor M_B.

Therefore, a current with same direction can pass through M_C.

O peration of the circuit for a positive input signal is described.
Operations of the transistors (on and off states) are clarified.

22

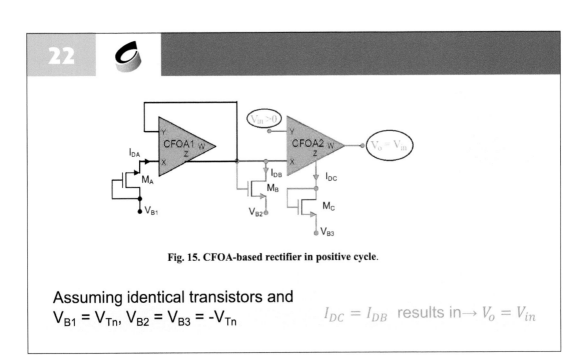

Fig. 15. CFOA-based rectifier in positive cycle.

Assuming identical transistors and
$V_{B1} = V_{Tn}, V_{B2} = V_{B3} = -V_{Tn}$

$I_{DC} = I_{DB}$ results in→ $V_o = V_{in}$

M atching conditions between voltages for proper operation of the circuit are discussed.
It is shown that output voltage is equal to input voltage for positive inputs.

23

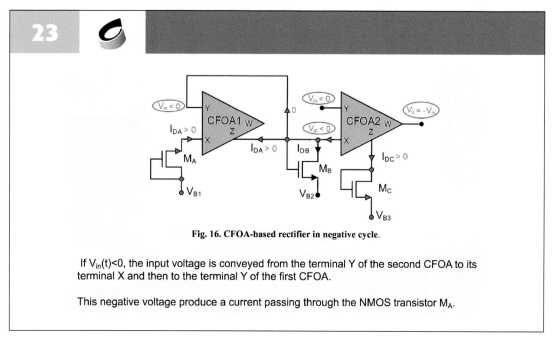

Fig. 16. CFOA-based rectifier in negative cycle.

If $V_{in}(t)<0$, the input voltage is conveyed from the terminal Y of the second CFOA to its terminal X and then to the terminal Y of the first CFOA.

This negative voltage produce a current passing through the NMOS transistor M_A.

O peration of the CFOA-based circuit for a negative input signal is described. On and off state transistors are determined.

24

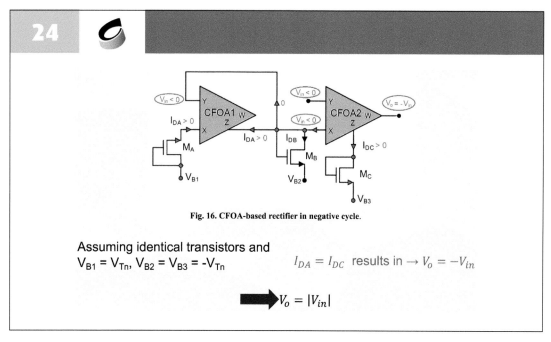

Fig. 16. CFOA-based rectifier in negative cycle.

Assuming identical transistors and
$V_{B1} = V_{Tn}$, $V_{B2} = V_{B3} = -V_{Tn}$

$I_{DA} = I_{DC}$ results in $\rightarrow V_o = -V_{in}$

➡ $V_o = |V_{in}|$

I t is shown that output voltage is equal to negative of the input voltage for negative input value signals. Overall output equation of the circuit is given.

SEC VI: CONCLUSION

25

✓ Conventional methotds of rectification are mentioned.

✓ Disadvantages of the conventional rectifier circuits are discussed.

✓ High precision methods of rectification are discussed.

✓ Several circuits using active elemnts such as CCII, OTA, etc, are given as examples.

C onclusion part of the presentation is given.
  Important points of the presentation are highlighted.

26   References

[1] R.C. Jaeger, T. N. Blalock "Microelectronic Circuit Design", 4th Edition, 2011. McGraw Hill Press.

[2] S. J. G. Gift, "A high-performance full-wave rectifier circuit," *Int. J. Electron.*, vol. 87, no. 8, pp. 925–930, 2008.

[3] M. Yildiz, S. Minaei, and E. Yuce, "A high-performance full-wave rectifier using a single CCII-, two diodes, and two resistors," *Sci. Iran. D*, vol. 24, no. 6, pp. 3280–3286, 2017.

[4] C. Psychalinos, S. Minaei, A. Yesil, "First-order inverse filters: Implementations using a single voltage conveyor and potential applications," Int. J. of Circuit Theory and Applications, vol. 50, no. 10, pp. 3704-3714, 2022.

[5] L. Safari, G. Barile, V. Stornelli, G. Ferri, "A new versatile full wave rectifier using voltage conveyors," AEU - International Journal of Electronics and Communications, vol. 122, p. 153267, 2020.

[6] M. Kumngern and K. Dejhan "High frequency and high precision CMOS full-wave rectifier," International Journal of Electronics, vol. 93, no. 3, pp. 185-199, 2006.

[7] E. Yuce, S. Minaei, M.A. Ibrahim "A novel full-wave rectifier/sinusoidal frequency doubler topology based on CFOAs, " Analog Integrated Circuits and Signal Processing, vol. 93, pp. 351-362, 2017.

T he references used in this presentation are listed.
  Some details on the references are given.

CHAPTER 02

Analog Integrated Filters Design Methodology

Prof. Costas Psychalinos

University of Patras, Physics Department,
Electronics Laboratory, Greece

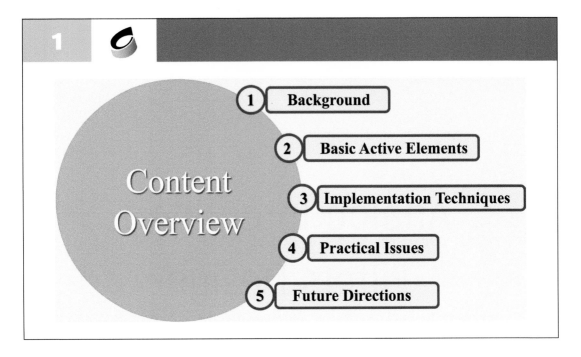

SEC I: BACKGROUND

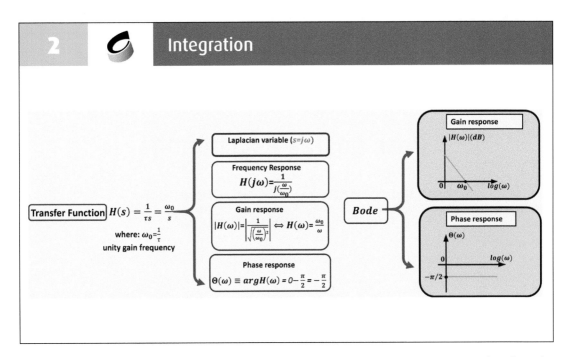

Integration is fundamental operation in analog signal processing. Integrators are employed in the implementation of filter stages, controllers, oscillators etc. Their unity gain frequency is determined by the associated time constant.

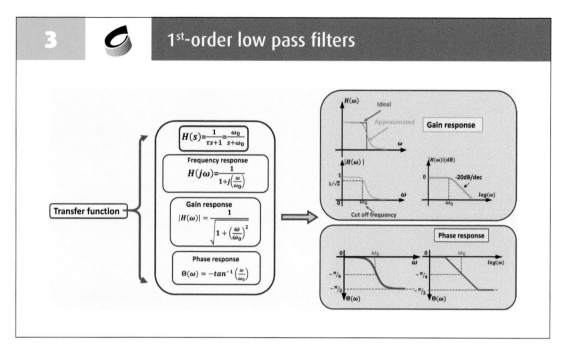

1st-order low pass filters are widely used in the construction of high-order filters. In this slide the basic properties of these filters in the frequency domain are provided, including the gain and phase responses. An important point, which plays important role in the compensation of operational amplifiers is that the maximum change of the phase which is implementable by these filters is equal to 90 deg.

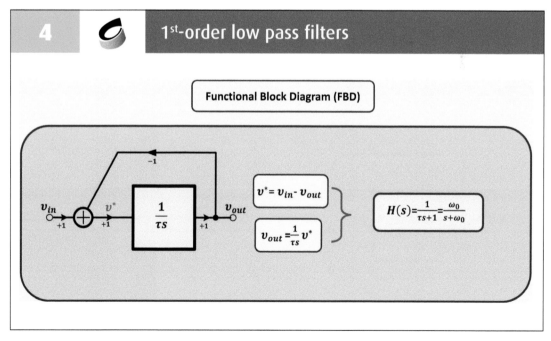

An 1st-order low pass filter can be realized using a lossless integrator in unity gain negative feedback loop.

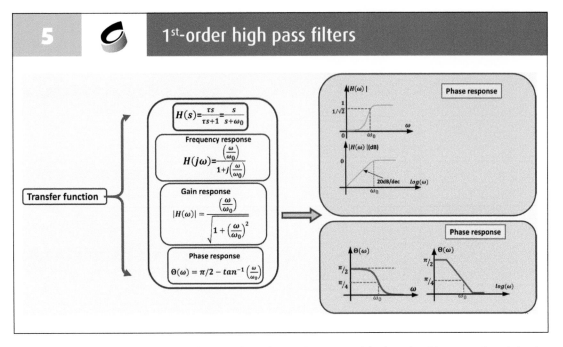

5 — **1ˢᵗ-order high pass filters**

1ˢᵗ-order high pass filters are widely used in the construction of high-order filters. In this slide the basic properties of these filters in the frequency domain are provided, including the gain and phase responses. Again, the maximum change of the phase which is implementable by these filters is equal to 90 deg.

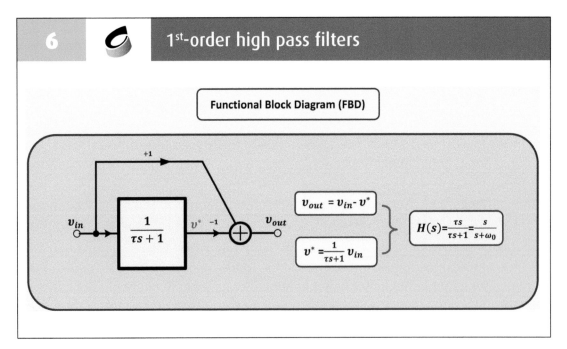

6 — **1ˢᵗ-order high pass filters**

An 1ˢᵗ-order high pass filter can be realized through the subtraction of the input signal and that produced by a 1ˢᵗ-order low pass filter. This offers modularity of the derivation of the basic 1ˢᵗ-order filter functions.

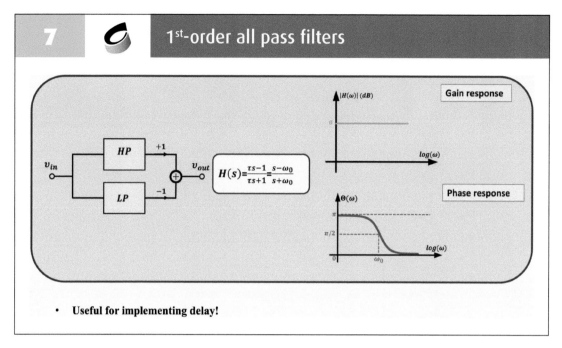

7 **1st-order all pass filters**

$$H(s)=\frac{\tau s-1}{\tau s+1}=\frac{s-\omega_0}{s+\omega_0}$$

- **Useful for implementing delay!**

All pass filters are useful for the implementation of delay stages and, therefore, adjusting the realized the group delay. They can be realized by subtracting the transfer functions of 1st-order low pass and high pass filter transfer functions.

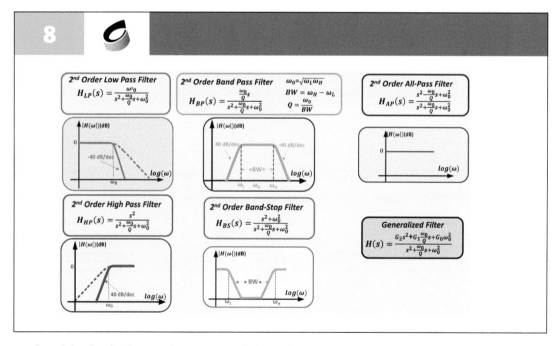

8

2nd Order Low Pass Filter

$$H_{LP}(s)=\frac{\omega_0^2}{s^2+\frac{\omega_0}{Q}s+\omega_0^2}$$

2nd Order Band Pass Filter $\quad \omega_0=\sqrt{\omega_L\omega_H}$

$$H_{BP}(s)=\frac{\frac{\omega_0}{Q}s}{s^2+\frac{\omega_0}{Q}s+\omega_0^2}$$

$$BW=\omega_H-\omega_L$$
$$Q=\frac{\omega_0}{BW}$$

2nd Order All-Pass Filter

$$H_{AP}(s)=\frac{s^2-\frac{\omega_0}{Q}s+\omega_0^2}{s^2+\frac{\omega_0}{Q}s+\omega_0^2}$$

2nd Order High Pass Filter

$$H_{HP}(s)=\frac{s^2}{s^2+\frac{\omega_0}{Q}s+\omega_0^2}$$

2nd Order Band-Stop Filter

$$H_{BS}(s)=\frac{s^2+\omega_0^2}{s^2+\frac{\omega_0}{Q}s+\omega_0^2}$$

Generalized Filter

$$H(s)=\frac{G_2s^2+G_1\frac{\omega_0}{Q}s+G_0\omega_0^2}{s^2+\frac{\omega_0}{Q}s+\omega_0^2}$$

In this slide the fundamental properties of the 2nd-order basic filter functions are presented. Using these basic filters, arbitrary 2nd-order filter function can be implemented through the employment of summation/subtraction of the basic filters functions.

SEC II: BASIC ACTIVE ELEMENTS

9 — Operational Amplifier (Op-Amp)

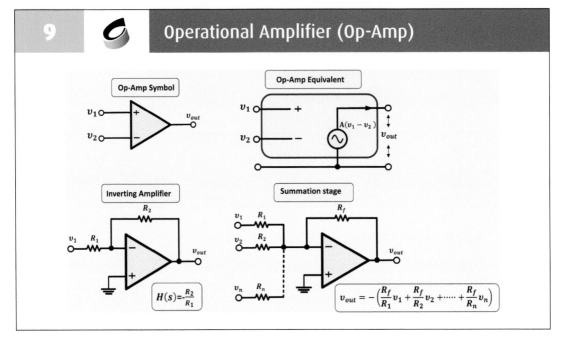

An ideal op-amp offers equalized voltages between its inputs, zero current at its inputs (i.e., infinite input impedance), and output voltage independent from the connected load (i.e., zero output resistance). Basic amplification and summation stages are provided in this slide.

10 — Basic signal processing stages using Op-Amps

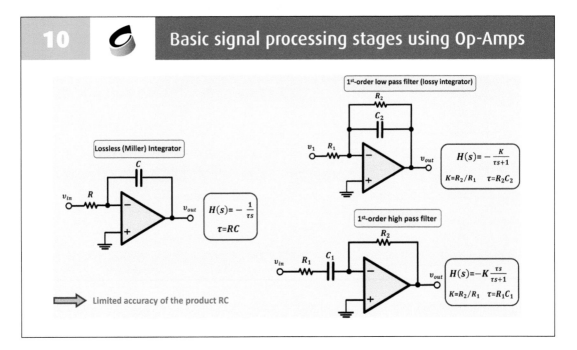

Lossless (Miller), lossy (1st-order low pass filter) integrator stages, as well as 1st-order high pass filter implementations are presented, where the maximum gain of the filters is determined by the ratio of 2 resistors and the time constant is formed by a RC product.

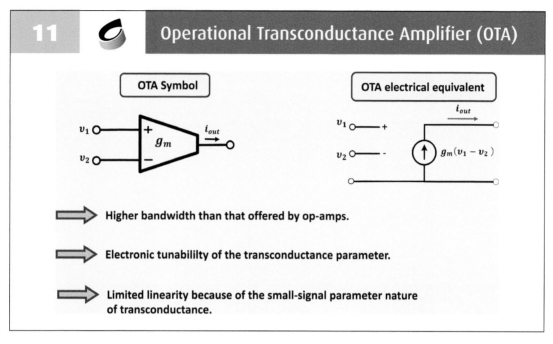

11 Operational Transconductance Amplifier (OTA)

OTA Symbol

OTA electrical equivalent

v_1, v_2, i_{out}, g_m

$g_m(v_1 - v_2)$

➡ Higher bandwidth than that offered by op-amps.

➡ Electronic tunabililty of the transconductance parameter.

➡ Limited linearity because of the small-signal parameter nature of transconductance.

Operational Transconductance Amplifier (OTA) is a fundamental element for performing analog signal processing. It is a voltage controlled current source with the transconductance parameter being the control variable. The input impedance is infinite (no current flow at the input terminals), and the output impedance is also infinite due to the fact that the output signal is current.

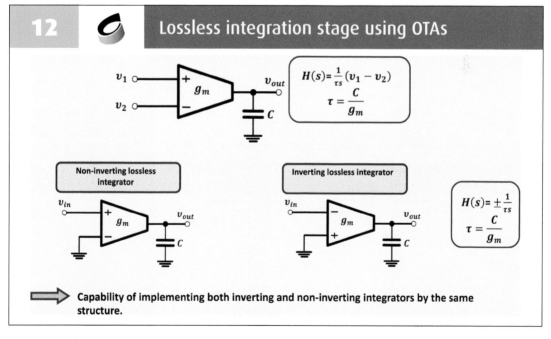

12 Lossless integration stage using OTAs

$$H(s) = \frac{1}{\tau s}(v_1 - v_2)$$
$$\tau = \frac{C}{g_m}$$

Non-inverting lossless integrator

Inverting lossless integrator

$$H(s) = \pm \frac{1}{\tau s}$$
$$\tau = \frac{C}{g_m}$$

➡ Capability of implementing both inverting and non-inverting integrators by the same structure.

OTA-C implementations of lossless integrator are presented in this slide. Owing to the differential nature of the input, the implementation of inverting and non-inverting transfer functions is capable just by grounding the appropriate input. This provides design versatility and flexibility.

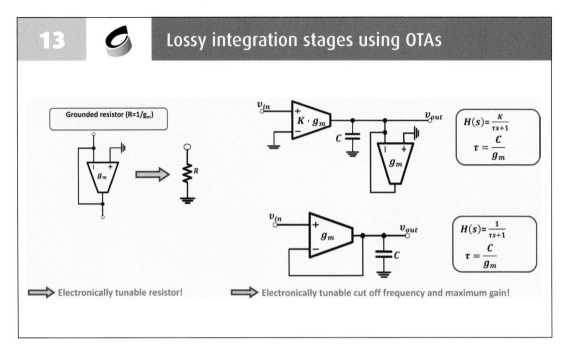

OTA-C implementations of lossless integration stages (i.e., 1st-order low pass stages) are presented in this slide. Thanks to the electronic adjustability of the small-signal parameter gm, the frequency characteristics of the filters can be tuned through the utilization of dc bias currents/voltages. In addition, the absence of resistors is a benefit from the required silicon area point of view.

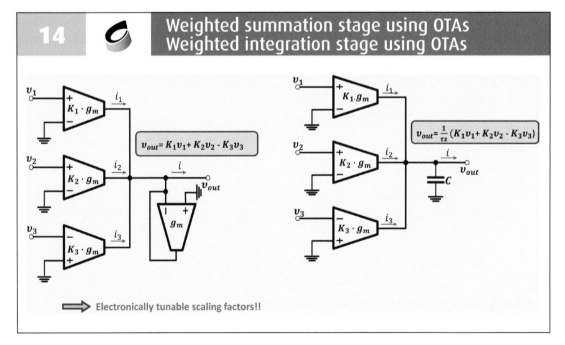

Pure summation and summation-integration OTA-C stages are presented. The attractive feature is that the scaling factors as well as the time-constant (in the case of integrator) can be electronically tubed by appropriate dc currents/voltages. In addition, the signs of summations are easily interchangeable due to the differential input of the OTAs.

SEC III: IMPLEMENTATION TECHNIQUES

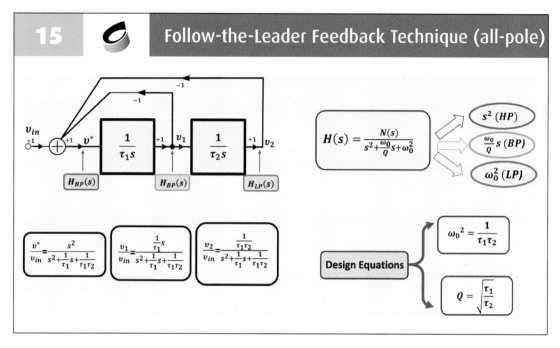

15 — **Follow-the-Leader Feedback Technique (all-pole)**

$$H(s) = \frac{N(s)}{s^2 + \frac{\omega_0}{Q}s + \omega_0^2}$$

s^2 (HP)

$\frac{\omega_0}{Q}s$ (BP)

ω_0^2 (LP)

$$\frac{v^*}{v_{in}} = \frac{s^2}{s^2 + \frac{1}{\tau_1}s + \frac{1}{\tau_1\tau_2}}$$

$$\frac{v_1}{v_{in}} = \frac{\frac{1}{\tau_1}s}{s^2 + \frac{1}{\tau_1}s + \frac{1}{\tau_1\tau_2}}$$

$$\frac{v_2}{v_{in}} = \frac{\frac{1}{\tau_1\tau_2}}{s^2 + \frac{1}{\tau_1}s + \frac{1}{\tau_1\tau_2}}$$

Design Equations

$$\omega_0^2 = \frac{1}{\tau_1\tau_2}$$

$$Q = \sqrt{\frac{\tau_1}{\tau_2}}$$

The Functional Block Diagram (FBD) of a Follow-the-Leader (FLF) multi-feedback structure is presented. This is capable for realizing the basic all-pole (without zeros) 2nd-order filter functions and they are simultaneously available at different points of the topology. The design equations are obtained by the equalization of the coefficients of the realized transfer functions and the generalized ones which describe the corresponding filters.

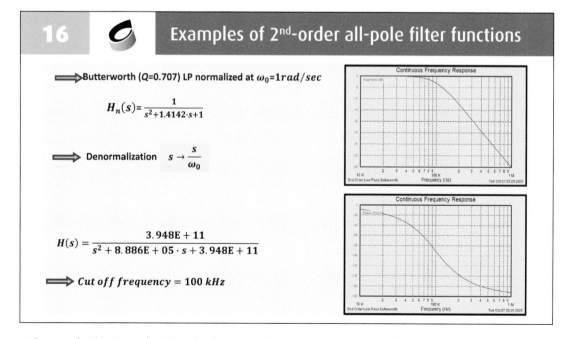

16 — **Examples of 2nd-order all-pole filter functions**

➡️ Butterworth (Q=0.707) LP normalized at $\omega_0 = 1 rad/sec$

$$H_n(s) = \frac{1}{s^2 + 1.4142 \cdot s + 1}$$

➡️ Denormalization $\quad s \rightarrow \frac{s}{\omega_0}$

$$H(s) = \frac{3.948E+11}{s^2 + 8.886E+05 \cdot s + 3.948E+11}$$

➡️ $Cut\ off\ frequency = 100\ kHz$

The transfer function of a 2nd-order low pass filter function is provided. The normalized (wo=1rad/sec) filter functions are available in Tables, and the designer must perform de-normalization according to the design specs.

17 — Examples of 2nd-order all-pole filter functions

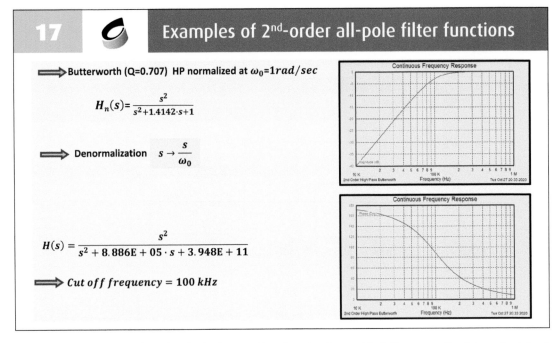

➡ **Butterworth (Q=0.707) HP normalized at $\omega_0=1rad/sec$**

$$H_n(s) = \frac{s^2}{s^2 + 1.4142 \cdot s + 1}$$

➡ **Denormalization** $\quad s \to \dfrac{s}{\omega_0}$

$$H(s) = \frac{s^2}{s^2 + 8.886E + 05 \cdot s + 3.948E + 11}$$

➡ *Cut off frequency = 100 kHz*

The transfer function of a 2nd order high pass filter function is provided. The normalized (wo=1rad/sec) filter functions are available in Tables, and the designer must perform de-normalization according to the design specs.

18 — OTA-C implementation of all-pole FLF

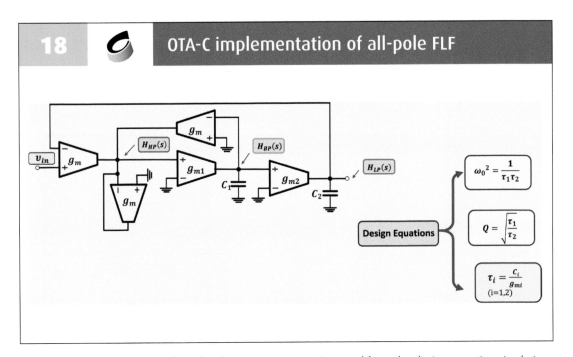

Design Equations

$$\omega_0{}^2 = \frac{1}{\tau_1 \tau_2}$$

$$Q = \sqrt{\frac{\tau_1}{\tau_2}}$$

$$\tau_i = \frac{C_i}{g_{mi}}$$
(i=1,2)

The OTA-C implementation of a 2nd-order FLF structure is given and from the design equations is obvious the electronic tuning capability of the wo and Q of the filter.

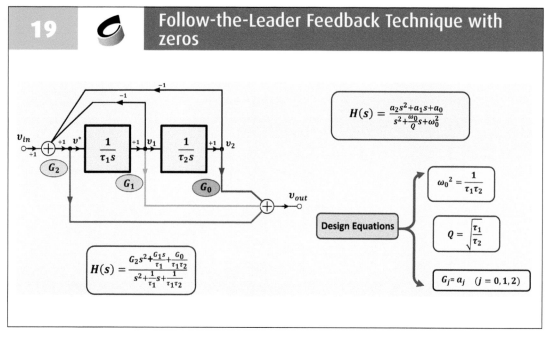

The FBD of a generalized 2nd-order transfer function is provided, which is derived from the all-pole FBD by adding algebraic summation of the intermediate outputs. The design equations are obtained by the equalization of the coefficients of the realized transfer function and the generalized one.

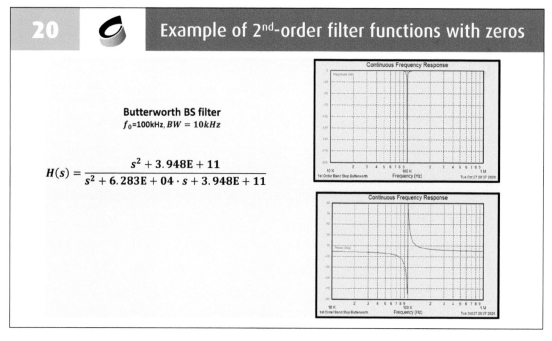

Example of a 2nd-order band stop (BS) filter transfer function, accompanied by the associated gain and phase responses.

21 · OTA-C implementation of FLF with zeros

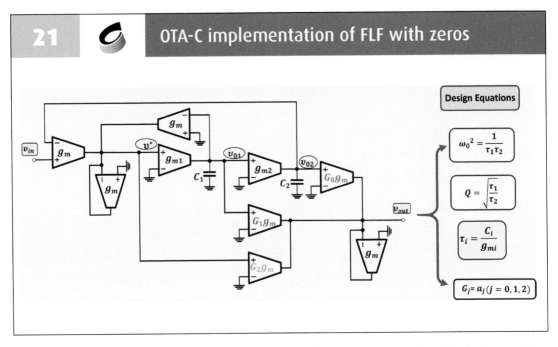

Design Equations

$$\omega_0{}^2 = \frac{1}{\tau_1 \tau_2}$$

$$Q = \sqrt{\frac{\tau_1}{\tau_2}}$$

$$\tau_i = \frac{C_i}{g_{mi}}$$

$$G_j = a_j \, (j = 0, 1, 2)$$

The OTA-C implementation of a 2nd-order FLF structure with zeros is given and from the design equations is obvious the electronic tuning capability of the characteristics of the filter.

22 · Generalized Follow-the-Leader Feedback Technique

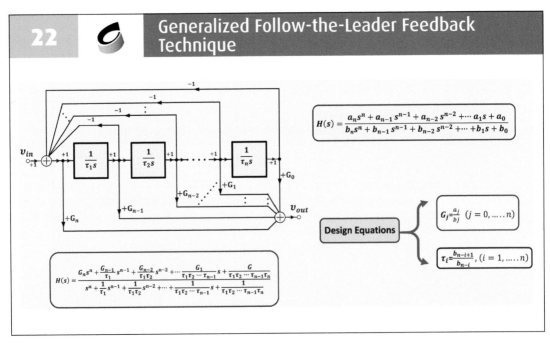

$$H(s) = \frac{a_n s^n + a_{n-1} s^{n-1} + a_{n-2} s^{n-2} + \cdots a_1 s + a_0}{b_n s^n + b_{n-1} s^{n-1} + b_{n-2} s^{n-2} + \cdots + b_1 s + b_0}$$

$$H(s) = \frac{G_n s^n + \frac{G_{n-1}}{\tau_1} s^{n-1} + \frac{G_{n-2}}{\tau_1 \tau_2} s^{n-2} + \cdots \frac{G_1}{\tau_1 \tau_2 \cdots \tau_{n-1}} s + \frac{G}{\tau_1 \tau_2 \cdots \tau_{n-1} \tau_n}}{s^n + \frac{1}{\tau_1} s^{n-1} + \frac{1}{\tau_1 \tau_2} s^{n-2} + \cdots \frac{1}{\tau_1 \tau_2 \cdots \tau_{n-1}} s + \frac{1}{\tau_1 \tau_2 \cdots \tau_{n-1} \tau_n}}$$

Design Equations

$$G_j = \frac{a_j}{b_j} \ (j = 0, \ldots . n)$$

$$\tau_i = \frac{b_{n-i+1}}{b_{n-i}}, \ (i = 1, \ldots . n)$$

FBD of a nth-order generalized filter function, accompanied by the associated design equations.

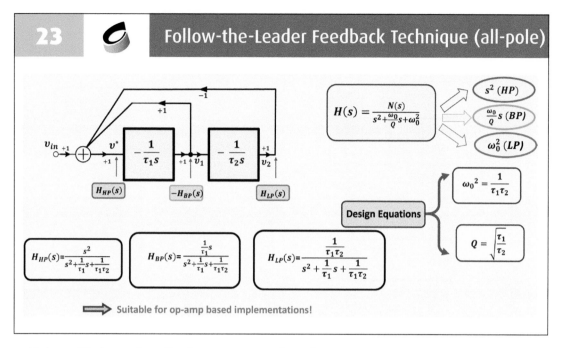

23 Follow-the-Leader Feedback Technique (all-pole)

$$H(s) = \frac{N(s)}{s^2 + \frac{\omega_0}{Q}s + \omega_0^2}$$

s^2 (HP)

$\frac{\omega_0}{Q}s$ (BP)

ω_0^2 (LP)

v_{in}

v^*

$-\frac{1}{\tau_1 s}$

$-\frac{1}{\tau_2 s}$

$H_{HP}(s)$

$-H_{BP}(s)$

$H_{LP}(s)$

Design Equations

$$\omega_0^2 = \frac{1}{\tau_1\tau_2}$$

$$Q = \sqrt{\frac{\tau_1}{\tau_2}}$$

$$H_{HP}(s) = \frac{s^2}{s^2 + \frac{1}{\tau_1}s + \frac{1}{\tau_1\tau_2}}$$

$$H_{BP}(s) = \frac{\frac{1}{\tau_1}s}{s^2 + \frac{1}{\tau_1}s + \frac{1}{\tau_1\tau_2}}$$

$$H_{LP}(s) = \frac{\frac{1}{\tau_1\tau_2}}{s^2 + \frac{1}{\tau_1}s + \frac{1}{\tau_1\tau_2}}$$

⟹ Suitable for op-amp based implementations!

This is modified FBD of an all-pole FLF structure, where the integration stages are inverting. The reason for doing this is that the transfer functions of integrators realized using op-amps as active elements are inverting. This means that extra inversion stages are required for realizing non-inverting integrator stages. The design equations are derived through the same way as in the case of the original FBD.

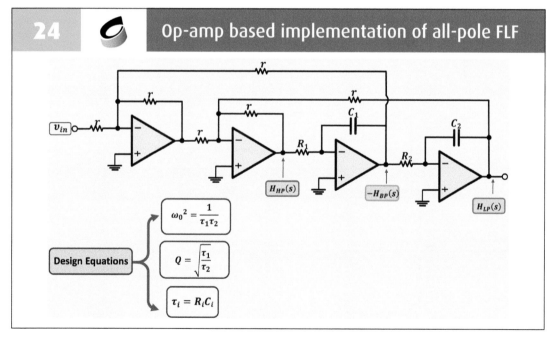

24 Op-amp based implementation of all-pole FLF

v_{in}

r

r

r

r

r

r

C_1

C_2

R_1

R_2

$H_{HP}(s)$

$-H_{BP}(s)$

$H_{LP}(s)$

Design Equations

$$\omega_0^2 = \frac{1}{\tau_1\tau_2}$$

$$Q = \sqrt{\frac{\tau_1}{\tau_2}}$$

$$\tau_i = R_i C_i$$

Implementation of the modified all-pole FBD using op-amps as active elements. The time constants are formed as resistor-capacitor products.

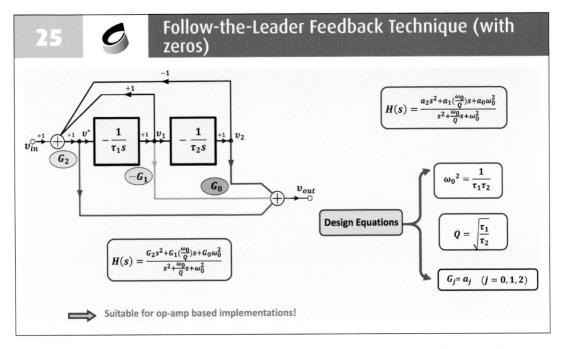

25 Follow-the-Leader Feedback Technique (with zeros)

$$H(s) = \frac{a_2 s^2 + a_1 (\frac{\omega_0}{Q}) s + a_0 \omega_0^2}{s^2 + \frac{\omega_0}{Q} s + \omega_0^2}$$

$$H(s) = \frac{G_2 s^2 + G_1 (\frac{\omega_0}{Q}) s + G_0 \omega_0^2}{s^2 + \frac{\omega_0}{Q} s + \omega_0^2}$$

Design Equations

$$\omega_0^2 = \frac{1}{\tau_1 \tau_2}$$

$$Q = \sqrt{\frac{\tau_1}{\tau_2}}$$

$$G_j = a_j \quad (j = 0, 1, 2)$$

Suitable for op-amp based implementations!

This is a modified FBD of an FLF structure, with inverting integration stages, suitable for implementing transfer function with zeros using op-amps as active elements.

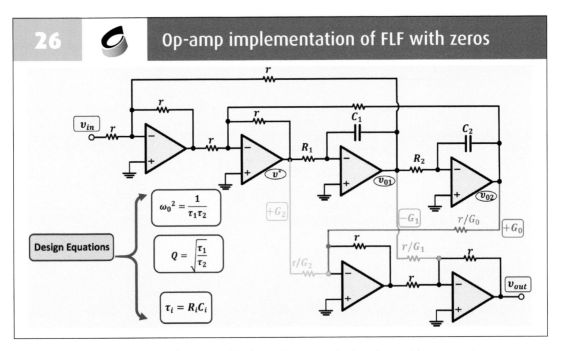

26 Op-amp implementation of FLF with zeros

Design Equations

$$\omega_0^2 = \frac{1}{\tau_1 \tau_2}$$

$$Q = \sqrt{\frac{\tau_1}{\tau_2}}$$

$$\tau_i = R_i C_i$$

Implementation of the modified FBD, suitable for realizing transfer functions with zeros, using op-amps as active elements. The time constants are formed as resistor-capacitor products.

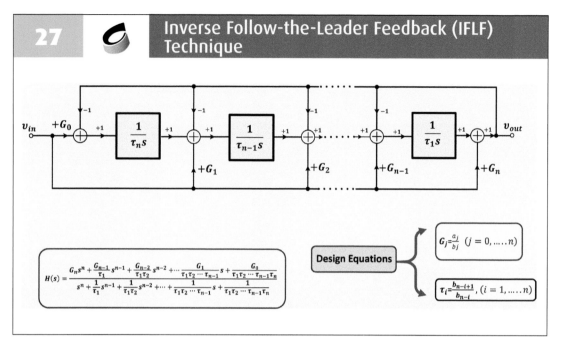

27 — Inverse Follow-the-Leader Feedback (IFLF) Technique

$$H(s) = \frac{G_n s^n + \frac{G_{n-1}}{\tau_1} s^{n-1} + \frac{G_{n-2}}{\tau_1 \tau_2} s^{n-2} + \cdots \frac{G_1}{\tau_1 \tau_2 \cdots \tau_{n-1}} s + \frac{G_0}{\tau_1 \tau_2 \cdots \tau_{n-1} \tau_n}}{s^n + \frac{1}{\tau_1} s^{n-1} + \frac{1}{\tau_1 \tau_2} s^{n-2} + \cdots + \frac{1}{\tau_1 \tau_2 \cdots \tau_{n-1}} s + \frac{1}{\tau_1 \tau_2 \cdots \tau_{n-1} \tau_n}}$$

Design Equations

$$G_j = \frac{a_j}{b_j} \quad (j = 0, \ldots . . n)$$

$$\tau_i = \frac{b_{n-i+1}}{b_{n-i}}, \quad (i = 1, \ldots . . n)$$

Another alternative for implementing arbitrary order filter functions. It is known as Inverse Follow-the-Leader Feedback structure, and it is useful in the case where active elements of differential input are utilized (such as OTAs).

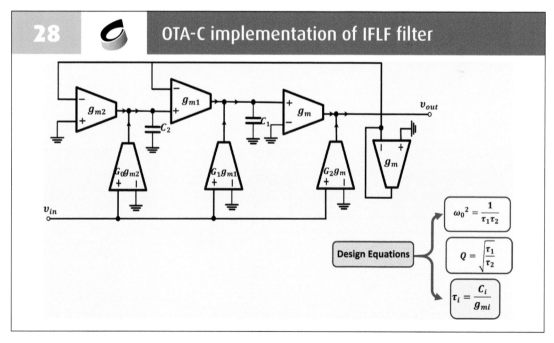

28 — OTA-C implementation of IFLF filter

Design Equations

$$\omega_0{}^2 = \frac{1}{\tau_1 \tau_2}$$

$$Q = \sqrt{\frac{\tau_1}{\tau_2}}$$

$$\tau_i = \frac{C_i}{g_{mi}}$$

OTA-C implementation of all-pole filter functions, based on the IFLF structure.

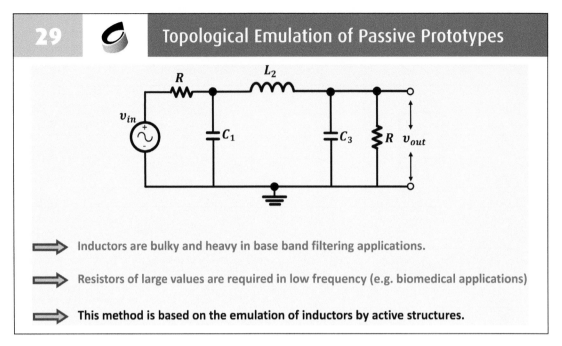

29 — **Topological Emulation of Passive Prototypes**

⟹ Inductors are bulky and heavy in base band filtering applications.

⟹ Resistors of large values are required in low frequency (e.g. biomedical applications)

⟹ **This method is based on the emulation of inductors by active structures.**

The previous material is oriented to the implementation of transfer functions. In this slide another alternative technique is presented, which is based on the emulation of the corresponding passive prototype filters. A way for doing this is the substitution of inductors by active structures. This is known as topological emulation of LCR prototypes.

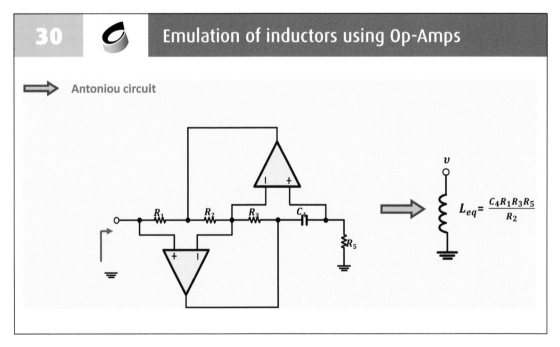

30 — **Emulation of inductors using Op-Amps**

⟹ Antoniou circuit

$$L_{eq} = \frac{C_4 R_1 R_3 R_5}{R_2}$$

The Generalized Impedance Converter (GIC) proposed by Antoniou is a powerful tool towards the emulation of an inductor. Choosing appropriate impedances in the general structure, the expression of the equivalent impedance is provided.

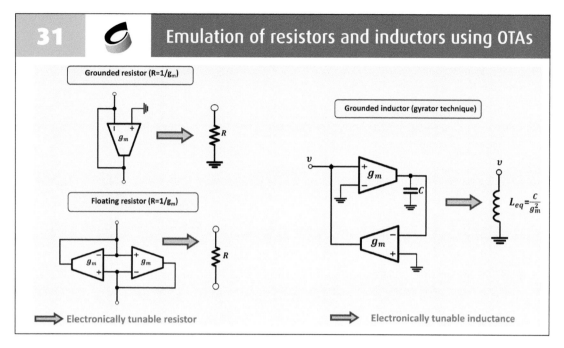

31 **Emulation of resistors and inductors using OTAs**

Grounded resistor (R=1/g_m)

Floating resistor (R=1/g_m)

Grounded inductor (gyrator technique)

$$L_{eq} = \frac{C}{g_m^2}$$

Electronically tunable resistor

Electronically tunable inductance

Possible implementations of electronically tunable resistors and inductor are provided, accompanied by the expressions about the equivalent resistance and inductance. The emulation of inductor is based on the gyration technique, where an integrator and a voltage-to-current (V/I) converter stages are utilized.

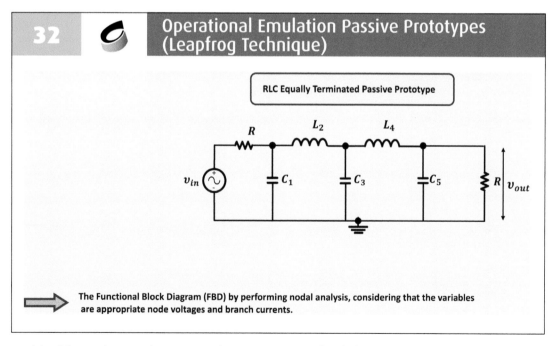

32 **Operational Emulation Passive Prototypes (Leapfrog Technique)**

RLC Equally Terminated Passive Prototype

The Functional Block Diagram (FBD) by performing nodal analysis, considering that the variables are appropriate node voltages and branch currents.

In this slide another an alternative technique is presented, which is based on the emulation of the operation of the corresponding passive prototype filters. This achieved by performing nodal analysis, considering that the variables are appropriate node voltages and branch currents.

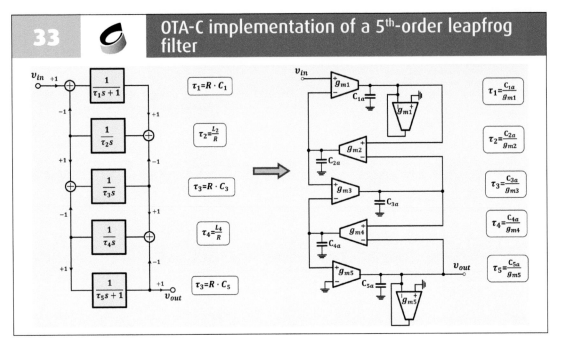

33 — OTA-C implementation of a 5th-order leapfrog filter

F BD derived from the passive prototype, by performing nodal analysis and considering that the variables are appropriate node voltages and branch currents. The corresponding OTA-C implementation is provided, along with the corresponding design equations.

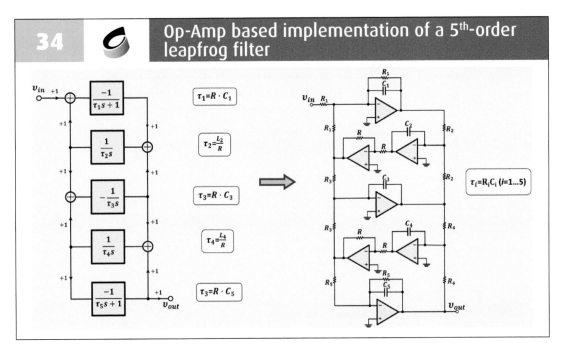

34 — Op-Amp based implementation of a 5th-order leapfrog filter

M odified FBD for facilitating the implantation using op-amps as active elements. The resulting op-amp RC implementation is given with the associated design equations.

SEC IV: PRACTICAL ISSUES

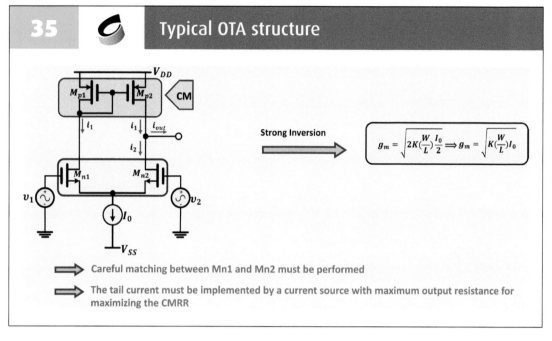

35 **Typical OTA structure**

Strong Inversion

$$g_m = \sqrt{2K(\frac{W}{L})\frac{I_0}{2}} \Longrightarrow g_m = \sqrt{K(\frac{W}{L})I_0}$$

Careful matching between Mn1 and Mn2 must be performed

The tail current must be implemented by a current source with maximum output resistance for maximizing the CMRR

A typical OTA structure is provided, with the corresponding expression of the transconductance parameter. In addition, some tips for improving the performance of the OTA are provided.

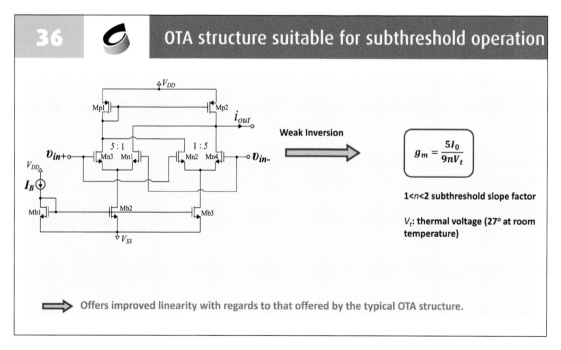

36 **OTA structure suitable for subthreshold operation**

Weak Inversion

$$g_m = \frac{5I_0}{9nV_t}$$

1<n<2 subthreshold slope factor

V_t: thermal voltage (27° at room temperature)

Offers improved linearity with regards to that offered by the typical OTA structure.

An OTA structure suitable for low-frequency applications is presented, which offers improved linear performance compared to the conventional implementation of the previous slide. The MOS transistors operate in the subthreshold region and this offers linear dependence of the gm parameter from the bias current instead of the square root dependence occurred in the case of the operation in the strong inversion region.

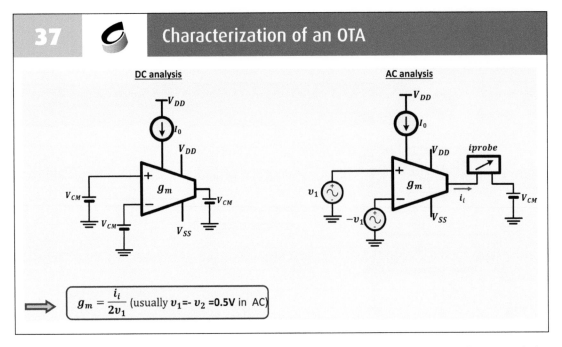

37 **Characterization of an OTA**

DC analysis

AC analysis

$$g_m = \frac{i_i}{2v_1} \text{ (usually } v_1 = -v_2 = 0.5V \text{ in AC)}$$

ips for evaluating the performance of an OTA using typical emulators (SPICE, Cadence etc.) are provided in this slide. The voltage Vcm is usually the mean value of the power supply voltage i.e., Vcm=(Vdd+Vss)/2.

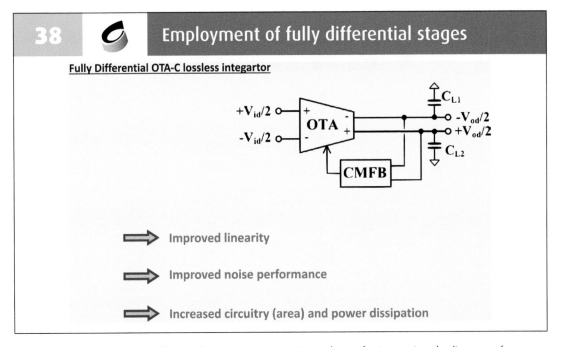

38 **Employment of fully differential stages**

Fully Differential OTA-C lossless integartor

Improved linearity

Improved noise performance

Increased circuitry (area) and power dissipation

he employment of fully differential stages is an attractive solution for improving the linear performance of OTA-C structures, because they suffer from the limited input range which can be efficiently handled by these active cells. The price paid is the increased area and power dissipation.

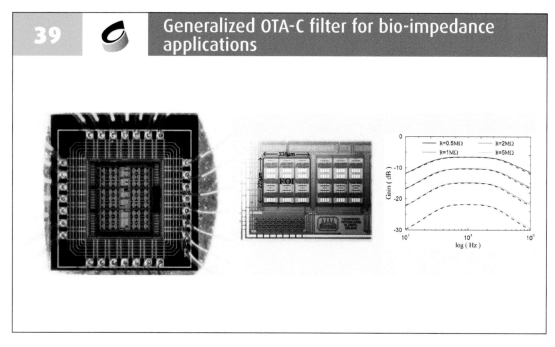

A n example of fabricated chip containing an OTA-C 2nd-order generalized filter is provided in this slide. This filter is used for implementing a fractional-order capacitor emulator and it was the first one in the literature.

SEC V: FUTURE DIRECTIONS

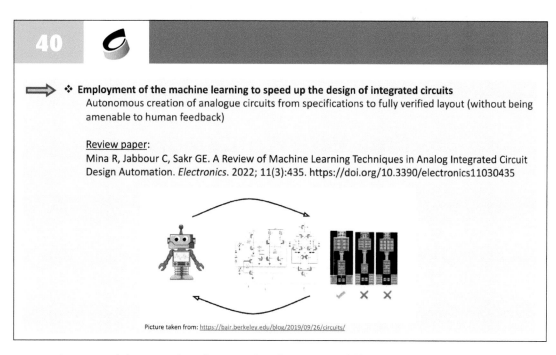

F uture directions of the research in the area of analog integrated filters are discussed in this slide.

Circuits and Systems for Machine Learning and Artificial Intelligence Applications

Prof. Paul P. Sotiriadis

National Technical University of Athens
School of Electrical & Computer Engineering, Greece

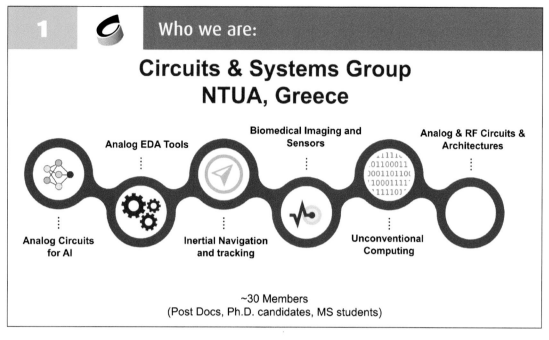

Circuits and Systems Group @ NTUA – Research Directions.

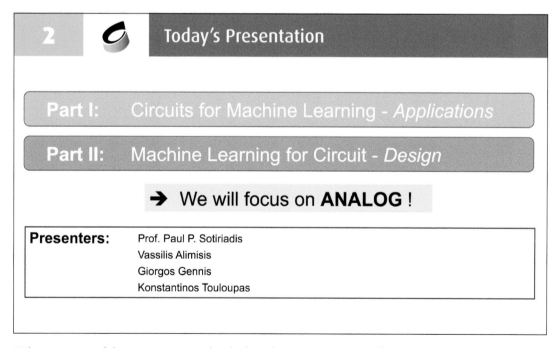

The two parts of the presentation – The duality of the Circuits and Machine Learning relationship.

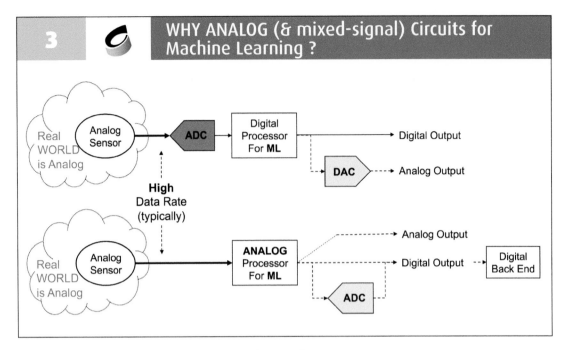

The Value of the Analog and the Mixed-Signal ML implementation.

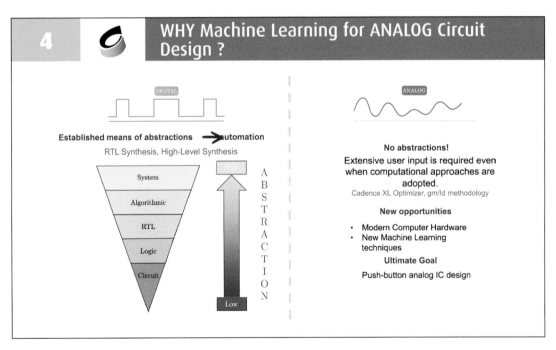

The lack of Analog Design Automation Tools (several efforts, minimal commercial products). The new opportunities based on ML and AI – driven approaches.

Part I Circuits for Machine Learning Applications

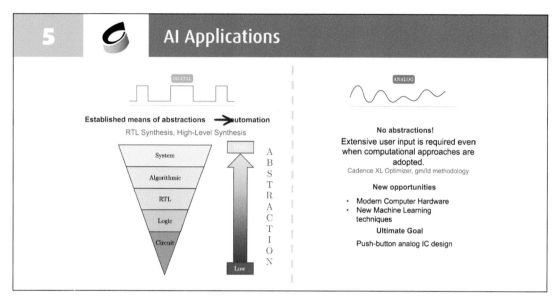

| 5 | AI Applications |

Machine Learning applications:

1. Biomedical Engineering: ML models are used to monitor the patient's condition or provide a diagnosis.
2. Robotics: Robotics systems and robotic artificial limbs depend on AI for navigation and control.
3. Computer Vision and Voice recognition: used to establish autonomous human-computer interfaces - That can identify objects and voice commands.
4. Automation in industrial environments and autonomous vehicles.

| 6 | ML/A.I. Hardware Processing |

HARDWARE	POWER	ACCURACY	Comments	Applications
Standard computer (CPU/ GPU) Software-based	HUGE	High	Requires Power hungry Data conversion Difficult portability	Big Data High performance
Dedicated Digital Circuit (FPGA)	High	High	Requires Power hungry Data conversion	Dedicated Big Data High performance
Analog Circuit	Low	Medium	More Challenging Design Based on transistors' characteristics	Low Power • Wearable • Implantable • Remote
Mixed-Signal Circuit	Medium	High	Certain advantages over purely digital and purely analog	Low Power • Portable • Wearable • Implantable • Remote

A comparison of Software/Hardware implementations of Machine Learning algorithms, with respect to power consumption, model's implementation accuracy and targeted applications. Analog and mixed-signal implementations provide an alternative to the standard, power hungry digital ones and are beneficial in applications where portability is required.

7 Classification System Chain

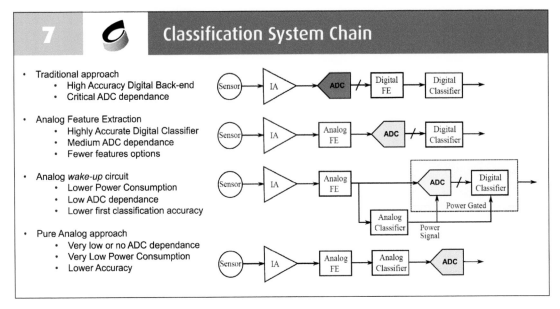

- Traditional approach
  - High Accuracy Digital Back-end
  - Critical ADC dependance

- Analog Feature Extraction
  - Highly Accurate Digital Classifier
  - Medium ADC dependance
  - Fewer features options

- Analog *wake-up* circuit
  - Lower Power Consumption
  - Low ADC dependance
  - Lower first classification accuracy

- Pure Analog approach
  - Very low or no ADC dependance
  - Very Low Power Consumption
  - Lower Accuracy

An intuitive illustration of a sensor acquisition system with a back-end classifier. Four distinct system level perspectives can be considered:

1. The traditional one, where Analog circuitry is used for interfacing the sensor only,
2. The "Analog Feature Extraction" one where the information to be fed to the classifier is extracted via Analog circuitry,
3. The "Analog Wake-up circuit" one where an Analog implementation of the Classifier is used to power up or down a more accurate Digital implementation
4. The purely Analog implementation where no Digital Classifier is involved.

The power requirements of the Analog to Digital Converter (ADC) are progressively reduced in each of the discussed perspectives.

8 Epileptic Seizure Prediction (wearable device)

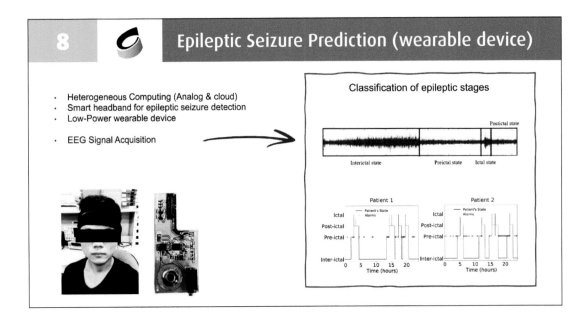

- Heterogeneous Computing (Analog & cloud)
- Smart headband for epileptic seizure detection
- Low-Power wearable device

- EEG Signal Acquisition

Classification of epileptic stages

An example of an Analog Classifier system implementation application: Epileptic seizure prediction headband. It is based on a heterogeneous computing ML, that combines Analog and Cloud computing. The device monitors the EEG signal and can predict an upcoming seizure, before it even occurs. The partial Analog-computing implementation greatly increases the battery life.

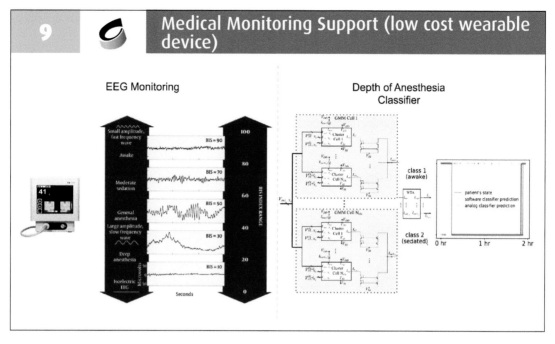

Analog ML application for low-cost wearable device: Binary Classification for determining whether a patient is sedated or awake. This enables informed decision making by the doctor, who decides whether to increase the anesthetic dosage or not.

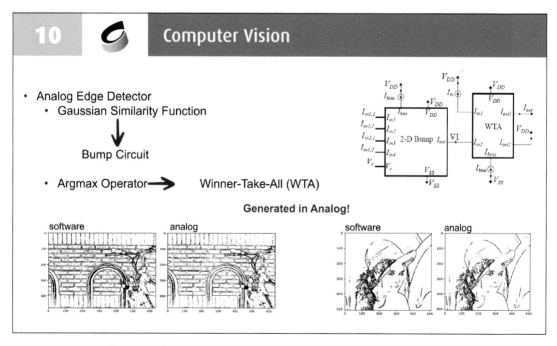

An example Analog AI implementation in computer vision: Edge Detection. An Edge Detection Analog cell derives the image gradient, centered at one pixel, to determine whether it is part of an edge or not. It consumes about 25nW, operating independently of other Edge Detector cells.

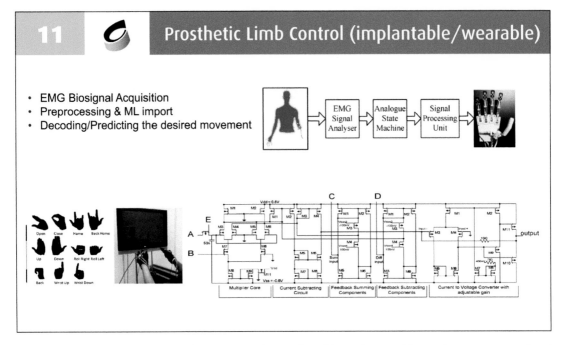

An example of Analog AI implementation for implantable and wearable applications: EMG signal classification. In the case of a partial limb amputation, EMG on the remaining limb part can still provide the necessary information to operate a prosthetic one.

An AI solution captures and processes surface EMG signals to control the prosthetic limb.

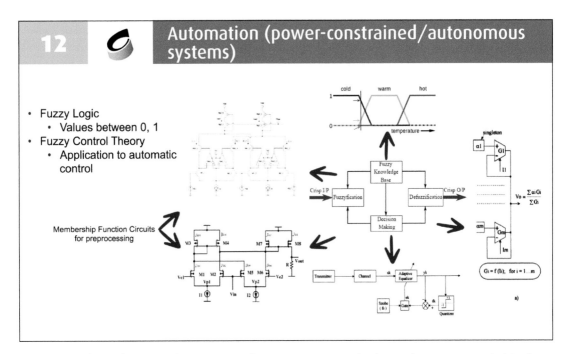

An example Analog AI implementation of a Fuzzy system: Crisp input data are converted to fuzzy data through the fuzzification step. Then the fuzzy decision is converted to a crisp decision through the defuzzification step.

Both these steps can be implemented using low power analog circuits, as the ones shown in this slide.

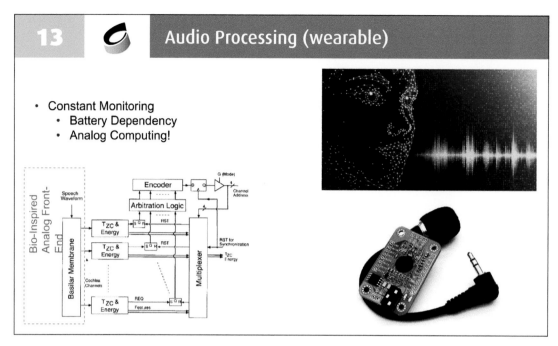

13 — Audio Processing (wearable)

- Constant Monitoring
  - Battery Dependency
  - Analog Computing!

An example of a mixed-signal AI voice recognition system. The analog front end is inspired by the human's Basilar Membrane to preprocess the raw input soundwaves.
The digital back-end receives the processed signal and provides the final decision.

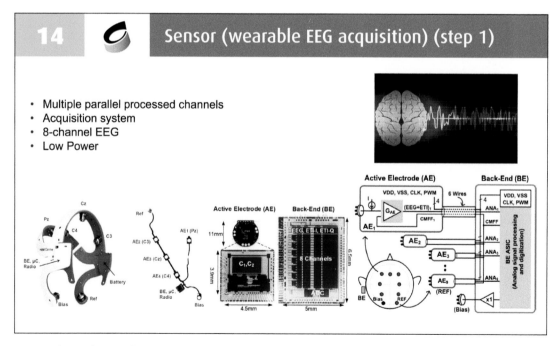

14 — Sensor (wearable EEG acquisition) (step 1)

- Multiple parallel processed channels
- Acquisition system
- 8-channel EEG
- Low Power

A purely Analog Machine Learning classification system with application to brain abnormality detection using EEG signals is described in the following slides.

The acquisition is performed using an active electrode-based EEG system.
The device is meant to be worn on the head and can process 8 different EEG channels in parallel.
Its Analog low-power part enables long battery life.

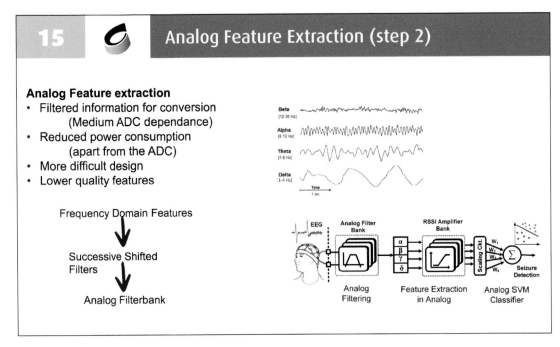

The pre-processing is performed by a Filter-based Analog Feature extraction system.

The acquired EEG signals are fed into 4 Filter banks driving a "received signal strength indicator" (RSSI) circuit.

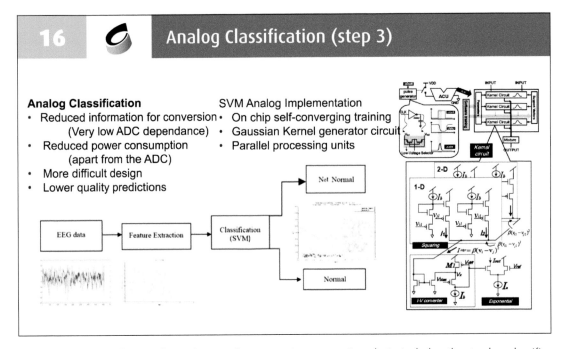

The final part of the purely Analog Machine Learning processing chain includes the Analog classifier implementation.

The classifier receives the Features from the Analog Feature extractor to detect patterns and abnormalities of the EEG signals.

The Analog implementation of the Support Vector Machine algorithm is highly parallelized and capable of on-chip training.

Part II Machine Learning for Circuit Design

17   Why automation?

A utomation in the field of Analog Integrated circuit Design is driven by two main factors:
First, the emergence of new electronic systems, the high-demand in high-end electronics and the tight time-to-market constraints.

Secondly, Digital design makes use of established automation procedures, while analog circuit design does not. This discrepancy must be addressed though the introduction of automation.

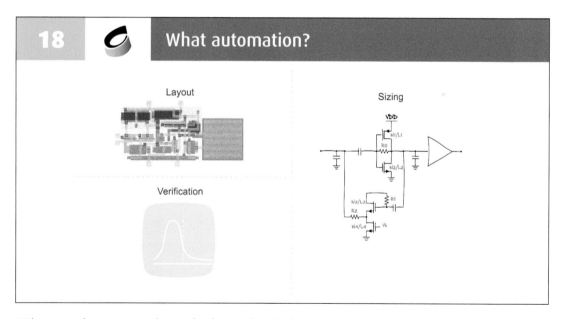

18   What automation?

T here are three main tasks involved in analog IC design where automation can be introduced using Machine Learning techniques:

Layout automation promises to automate the placement and the routing procedures of layout design, given an initial sized netlist.

Sizing automation promises to determine the geometric sizes of individual devices of a circuit provided a set of specifications.

Automation in the verification procedure of the Analog design cycle reduces the time to verify the operation of the circuit under test.

The automation in all three of these tasks is enabled through Machine Learning techniques.

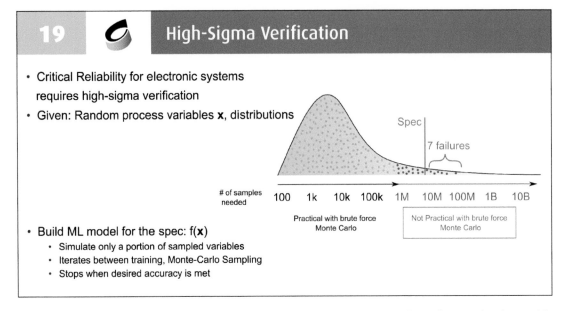

19 — High-Sigma Verification

- Critical Reliability for electronic systems requires high-sigma verification
- Given: Random process variables **x**, distributions

Spec

7 failures

# of samples needed — 100   1k   10k   100k   1M   10M   100M   1B   10B

Practical with brute force Monte Carlo   Not Practical with brute force Monte Carlo

- Build ML model for the spec: f(**x**)
  - Simulate only a portion of sampled variables
  - Iterates between training, Monte-Carlo Sampling
  - Stops when desired accuracy is met

Many circuit blocks, especially ones used in many multiples, like memory bit-cells, must be thoroughly verified. High-sigma verification, however, may require billions of simulations using brute force Monte Carlo (MC). Proprietary software by Solido (Siemens) uses ML to accelerate this procedure.

By using a number of initial MC samples, it learns to reproduce combinations of process-variable vectors resulting in circuit performance near the specification threshold; thus providing a way to achieve high-sigma verification with fewer simulations.

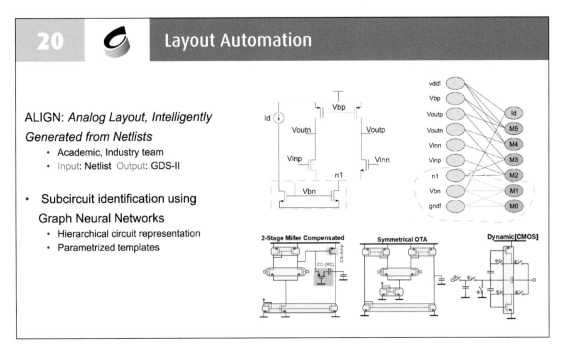

20 — Layout Automation

ALIGN: *Analog Layout, Intelligently Generated from Netlists*
  - Academic, Industry team
  - Input: **Netlist** Output: GDS-II

- Subcircuit identification using Graph Neural Networks
  - Hierarchical circuit representation
  - Parametrized templates

2-Stage Miller Compensated   Symmetrical OTA   Dynamic[CMOS]

Analog layout automation assumes input netlists, as well as a set of design rules, which often depend on the PDK used. The ALIGN suite is a set of tools achieving such automation.

ALIGN's main step in the automation is the definition of a hierarchy, according to which individual devices or basic building blocks are placed. In order to identify the individual building blocks, ALIGN uses Graph Neural Networks, which are trained using a large dataset of annotated netlists. The lowest levels of hierarchy are created by using pre-existing parametric templates.

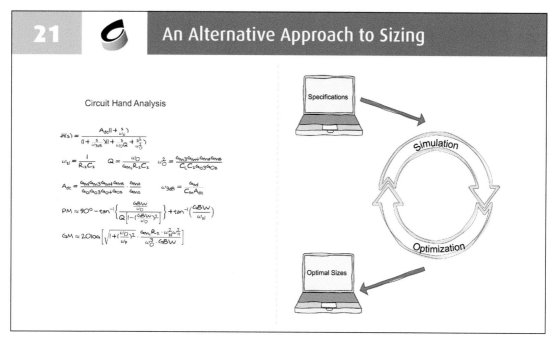

nstead of using closed-form Equations and closed-form circuit performance Metrics, sizing automation assumes a Simulation-in-The-Loop approach, where, given a set of desired specifications, a computer program repeatedly searches for the device sizes that optimally satisfy these specifications.

Of course, this iterative procedure consists of individual steps, which are depicted in the rightmost figure.

The evaluation engine is, most of the times, a commercial simulator using the PDK models.

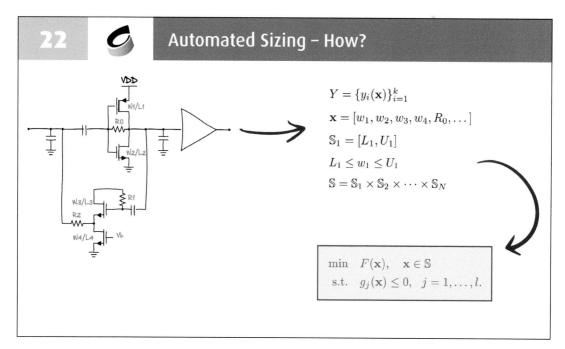

The automation of the sizing begins by considering a particular topology, such as the Low Noise Amplifier shown in this slide.

Also, the designer should determine which of the devices are parametrized and what variables are considered.

23   Automated Sizing – How?

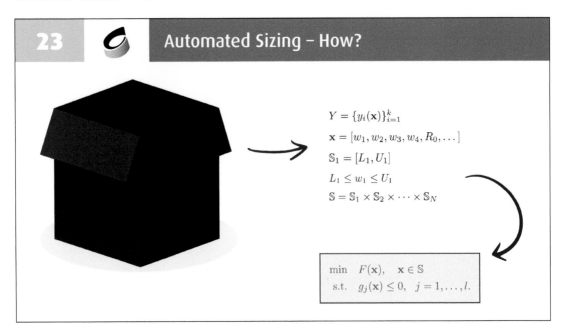

$$Y = \{y_i(\mathbf{x})\}_{i=1}^{k}$$
$$\mathbf{x} = [w_1, w_2, w_3, w_4, R_0, \dots]$$
$$\mathbb{S}_1 = [L_1, U_1]$$
$$L_1 \leq w_1 \leq U_1$$
$$\mathbb{S} = \mathbb{S}_1 \times \mathbb{S}_2 \times \cdots \times \mathbb{S}_N$$

$$\min \quad F(\mathbf{x}), \quad \mathbf{x} \in \mathbb{S}$$
$$\text{s.t.} \quad g_j(\mathbf{x}) \leq 0, \quad j = 1, \dots, l.$$

Using the aforementioned information, along with the ranges of the variables, a black-box optimization problem is formulated.

The set Y denotes the performance metrics that the designer examines.

Functions F and g are the objective and the constraints of the optimization, respectively, and they are determined by the performance metrics accounted for by the designer, and are evaluated using simulator's outputs.

24   Bayesian Optimization

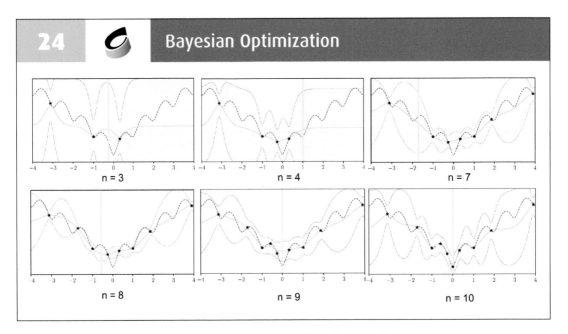

The solution of the optimization problem can be found by adopting a hyperparameter tuning approach and using Bayesian Optimization (BO).

BO fits Gaussian Process models to the already evaluated data and uses their uncertainty estimates, like the lower confidence bound in the depicted plots, to determine the location(s) of the following point(s) for evaluation.

BO typically requires few evaluations to reach a good approximation of the global optima, which is very important in the case of Analog circuit sizing since simulations may take considerable time to complete.

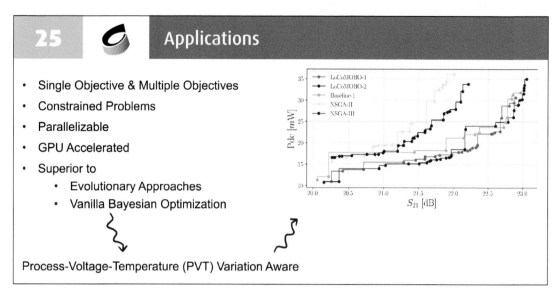

25   Applications

- Single Objective & Multiple Objectives
- Constrained Problems
- Parallelizable
- GPU Accelerated
- Superior to
  - Evolutionary Approaches
  - Vanilla Bayesian Optimization

Process-Voltage-Temperature (PVT) Variation Aware

Bayesian Optimization can be used in both Single Objective and Multi-Objective formulations, where the goal is to explore the design space of the given topology.

Also, modifications to its procedure make it possible for BO to handle constrained problems, provide multiple points for evaluation per iteration, which allows for parallel simulation execution, and can use GPUs to train its Gaussian Process models.

Within a minimal time budget, the LoCoMOBO implementation of BO can provide variation-aware designs for a single-ended LNA combining 45 corners, with process, voltage and temperature variations accounted for.

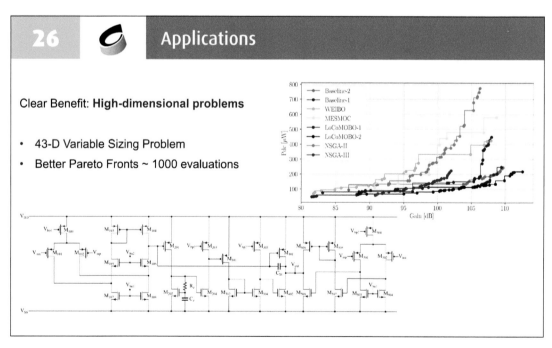

26   Applications

Clear Benefit: **High-dimensional problems**

- 43-D Variable Sizing Problem
- Better Pareto Fronts ~ 1000 evaluations

LoCoMOBO proves scalable to high-dimensional problems as well.

In this example, a 4 stage amplifier shown in the bottom of the slide is considered. This circuit is parametrized using 43 different variables, and LoCoMOBO results in better pareto fronts than many other AI black-box optimization alternatives.

In roughly 1000 simulations, LoCoMOBO provides very good trade offs between the DC gain and the power consumption of the examined circuit, with multiple constraints being fulfilled as well.

27 References

Circuits for Machine Learning Applications

- Lin, Shih-Kai, et al. "An ultra-low power smart headband for real-time epileptic seizure detection." *IEEE journal of translational engineering in health and medicine* 6 (2018): 1-10.
- Alimisis, V., Mouzakis, V., Gennis, G., Tsouvalas, E., & Sotiriadis, P. P. An Analog Nearest Class with Multiple Centroids Classifier Implementation, for Depth of Anesthesia Monitoring. *IEEE International Conference on Smart Systems and Power Management* (IC2SPM 2022)
- Gennis, Georgios, et al. "A general purpose, low power, analog integrated image edge detector, based on a current-mode Gaussian function circuit." Analog Integrated Circuits and Signal Processing (2022): 1-12.
- Mangieri, E. et al. "A novel analogue circuit for controlling prosthetic hands". IEEE Biomedical Circuits and Systems Conference 2008 (pp. 81-84).
- Dualibe, Carlos, Michel Verleysen, and P. Jespers. Design of analog fuzzy logic controllers in CMOS technologies: implementation, test and application. Springer Science & Business Media, 2007.
- Kumar, Nagendra, et al. "An analog VLSI chip with asynchronous interface for auditory feature extraction." IEEE Transactions on Circuits and Systems II: Analog and Digital Signal Processing 45.5 (1998): 600-606.
- Xu, Jiawei, et al. "A wearable 8-channel active-electrode EEG/ETI acquisition system for body area networks." IEEE Journal of Solid-State Circuits 49.9 (2014): 2005-2016.
- Zhang, Y. et al. "RSSI amplifier design for a feature extraction technique to detect seizures with analog computing" 2020 IEEE international symposium on circuits and systems (ISCAS). IEEE, 2020.
- Zhang, Renyuan, et al. "Design of programmable analog calculation unit by implementing support vector regression for approximate computing." IEEE Micro 38.6 (2018): 73-82.

Machine Learning for Circuit Design

- Kunal, Kishor, et al. "GANA: Graph convolutional network based automated netlist annotation for analog circuits." 2020 Design, Automation & Test in Europe Conference & Exhibition (DATE). IEEE, 2020.
- Kunal, Kishor, et al. "ALIGN: Open-source analog layout automation from the ground up." Proceedings of the 56th Annual Design Automation Conference 2019. 2019.
- Touloupas, Konstantinos, Nikos Chouridis, and Paul P. Sotiriadis. "Local Bayesian Optimization For Analog Circuit Sizing." 2021 58th ACM/IEEE Design Automation Conference (DAC). IEEE, 2021.
- Touloupas, Konstantinos, and Paul P. Sotiriadis. "LoCoMOBO: A local constrained multi-objective Bayesian optimization for analog circuit sizing." IEEE Transactions on Computer-Aided Design of Integrated Circuits and Systems (2021).
- McConaghy, Trent, et al. "High-Sigma Verification and Design." Variation-Aware Design of Custom Integrated Circuits: A Hands-on Field Guide (2013): 115-167.
- Siemens: https://static.sw.cdn.siemens.com/siemens-disw-assets/public/82966/en-US/Siemens-SW-Automotive-IC-design-demands-next-WP-81734%20-C1

CHAPTER 04

Machine Learning Classification on Printed Circuits

Prof. Georgios Zervakis

University of Patras, Department of Computer
Engineering & Informatics, Greece

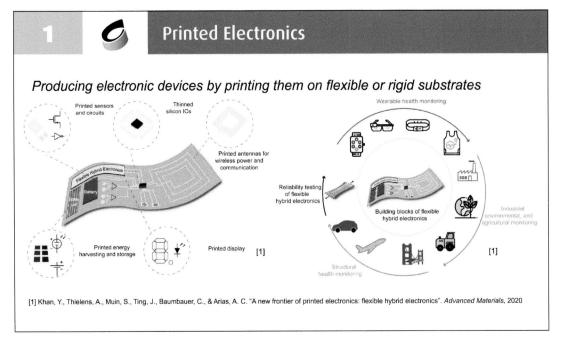

Printed Electronics

Producing electronic devices by printing them on flexible or rigid substrates

[1] Khan, Y., Thielens, A., Muin, S., Ting, J., Baumbauer, C., & Arias, A. C. "A new frontier of printed electronics: flexible hybrid electronics". *Advanced Materials*, 2020

Printed electronics denotes a set of printing methods which can realize ultra low-cost, large area and flexible computing systems.

Therefore, printed electronics emerge as a promising solution for application domains such as smart packaging, disposables, fast moving consumer goods, in-situ monitoring, and low-end healthcare products, like smart bandages.

Such domains pose requirements for ultra-low cost and conformality that silicon-based systems cannot satisfy.

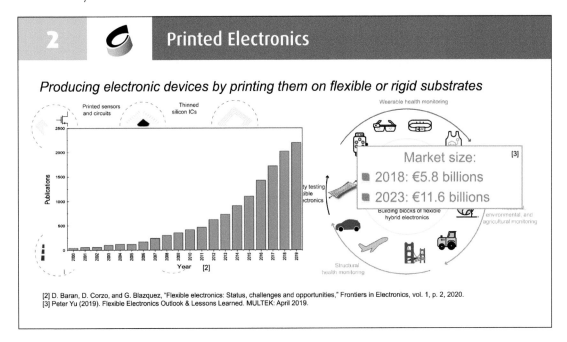

Printed Electronics

Producing electronic devices by printing them on flexible or rigid substrates

Market size: [3]
2018: €5.8 billions
2023: €11.6 billions

[2] D. Baran, D. Corzo, and G. Blazquez, "Flexible electronics: Status, challenges and opportunities," Frontiers in Electronics, vol. 1, p. 2, 2020.
[3] Peter Yu (2019). Flexible Electronics Outlook & Lessons Learned. MULTEK: April 2019.

As a result, due to these appealing features of printed electronics we are witnessing a vast growth in both printed electronics business and research.

For example, the market size of printed electronics in 2018 was 5.8 billion euros while it is expected to rise to almost 12 billions in 2023.

Similarly, the number of publications on printed electronics doubled from 2015 to 2019.

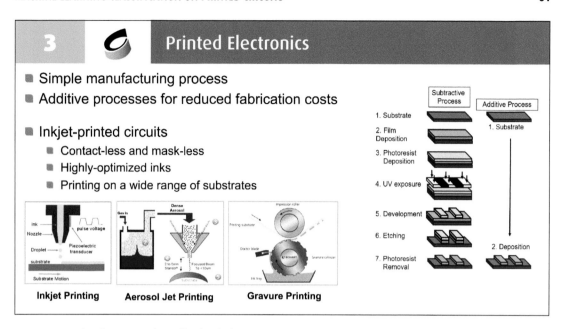

Printing technologies are broadly divided into two categories. Some printing technologies are based on purely additive manufacturing process, while others employ subtractive process as well. In the subtractive process, a series of deposition and etching steps are involved, similar to the silicon based processing.

Printed technologies include processes such as jet-printing, screen- or gravure-printing. Due to the simple manufacturing process as well as low equipment costs, ultra low-cost electronic circuits can be fabricated, at drastically lower cost compared to silicon-based processes that require expensive foundries and clean rooms. Several printing technologies can also be portable which reduces cost even further since a circuit can be manufactured on-demand at the point-of-use.

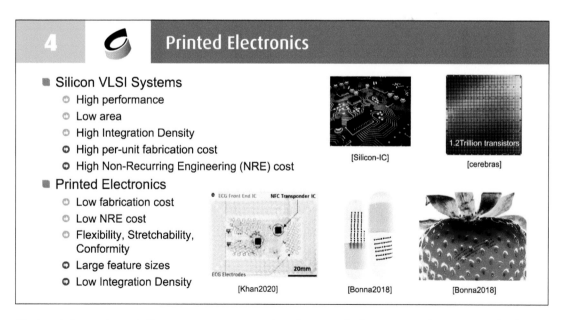

Printed electronics and silicon VLSI are not competitive but actually they are complementary technologies. Printed electronics will never challenge the performance and integration density of silicon systems.

On the other hand, in addition to the extremely low costs, printed electronics can also be flexible, stretchable, porous, and non-toxic.

These attributes make printed electronics even better fit for several disposable ultra-low-cost applications like for example smart packaging and on-body patches.

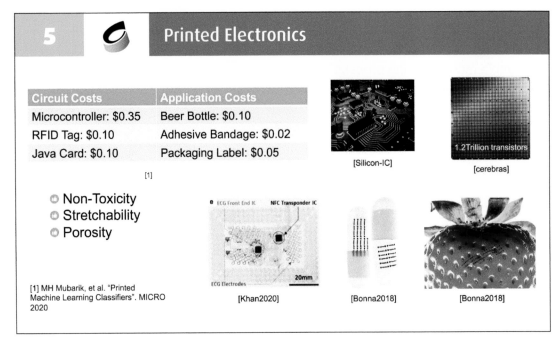

Such application domains are untouchable by silicon systems. Let's consider the cost requirement.
As an indicative example we present this table.
A silicon microprocessor can cost around 35 cents while the cost of the adhesive bandage is only 2 cents.
This example highlights that the costs of silicon systems are prohibitive for these applications.

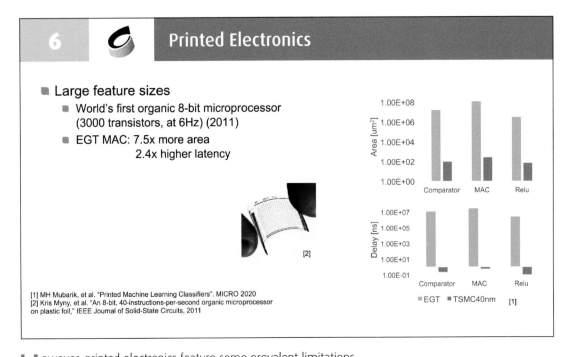

However, printed electronics feature some prevalent limitations.
The large feature sizes in printed electronics hinder the realization of complex printed systems.
For example, the first organic microprocessor had only 3000 transistors while an electrolyte-gated transistor MAC unit features 8 times higher area than the respective MAC unit at the TSMC40nm technology.

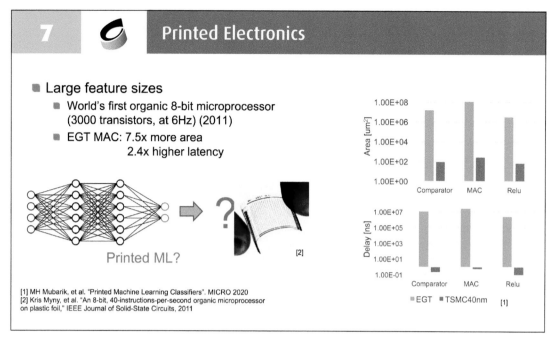

The figure on the right shows the area and delay of the printed comparator, MAC, and RELU units. As shown, the printed components feature orders of magnitude higher area than the corresponding silicon-based ones.

These components constitute the fundamental building blocks in typical machine learning classifiers and as a result, the realization of printed ML circuits becomes even more challenging, if feasible at all.

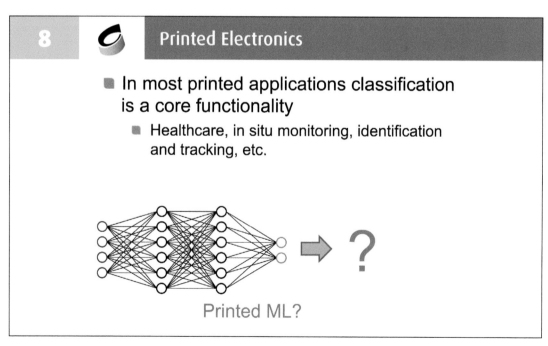

But why printed ML? Because the core task in most printed applications is classification.

9 Printed Electronics

- In most printed applications classification is a core functionality
  - Healthcare, in situ monitoring, identification and tracking, etc.

 ⇨ ?

Printed ML?

Is my beverage at the desired temperature?

src: sanpellegrino

Questions like: Is my beverage at the desired temperature?

10 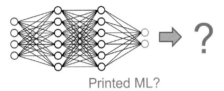 Printed Electronics

- In most printed applications classification is a core functionality
  - Healthcare, in situ monitoring, identification and tracking, etc.

⇨ ?

Printed ML?

Are the tomatoes ready to harvest?

src: wikipedia.org

Are the tomatoes ready to harvest?

11 **Printed Electronics**

- In most printed applications classification is a core functionality
  - Healthcare, in situ monitoring, identification and tracking, etc.

Is my wound healed?

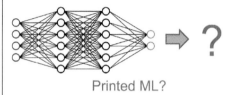 ➡ ?

Printed ML?

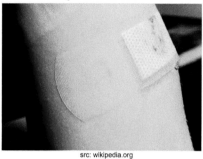

src: wikipedia.org

s my wound healed?

 12 **Printed Electronics**

- In most printed applications classification is a core functionality
  - Healthcare, in situ monitoring, identification and tracking, etc.

Is the milk still good?

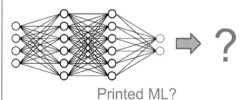 ➡ ?

Printed ML?

src: wikipedia.org

Or Is the milk still good?
Could be easily answered by a printed ML classifier.

13 Printed Electronics

■ In most printed applications classification
is a core functionality
  ■ Healthcare, in situ monitoring, identification
    and tracking, etc.

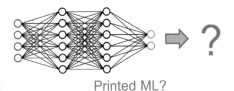

Printed ML?

■ First systematic approach for printed ML
classification in MICRO 2020 [1]
  ■ MLPs, SVM-C (SVM classification)

[1] MH Mubarik, et al. "Printed Machine Learning Classifiers". MICRO 2020

First systematic approach for printed ML classification is reported in MICRO 2020 where the authors used Electrolyte gated transistors and examined several ML classifiers.

14 Printed Electronics

■ In most printed applications classification
is a core functionality
  ■ Healthcare, in situ monitoring, identification
    and tracking, etc.

Printed ML?

■ First systematic approach for printed ML
classification in MICRO 2020 [1]
  ■ MLPs, SVM-C (SVM classification)

→ Area overhead: 21 to 2250 cm^2
Area in silicon: 0.004 to 0.51 mm^2

[1] MH Mubarik , DD Weller, N Bleier,M Tomei, J Aghassi-Hagmann, M Tahoori, R Kumar, "Printed Machine Learning Classifiers". MICRO 2020

The area overheads where excessive.

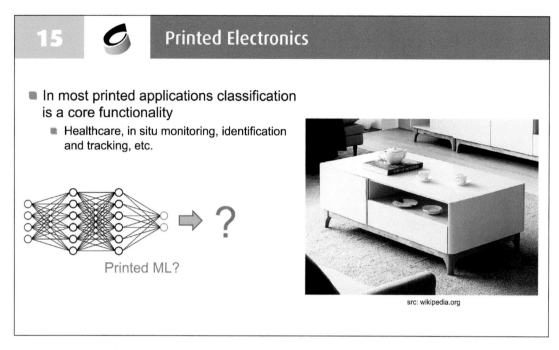

To put these numbers into perspective, the area of some classifiers reached the size of a coffee table.

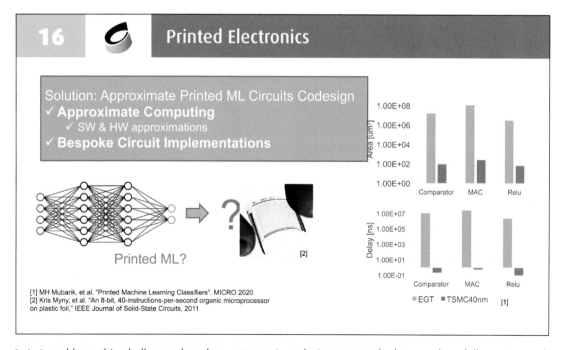

We address this challenge though a systematic codesign approach that employs fully customized, bespoke architectures enhanced by approximate computing at both the software and hardware levels.

17 Approximate Computing

- Approximate Computing is based on the fact that we demand too much accuracy from our computing systems
- Many applications exhibit an inherent error resilience
- Tradeoff little computational accuracy with significant area/power savings
- **Approximate Computing & Machine Learning form a perfect match**

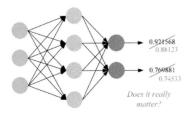

0.92**1568**
0.88123

0.76**9881**
0.74533

Does it really matter?

Task: Division

$$is \; \frac{923}{21} > 1.75?$$

$$is \; \frac{923}{21} > 45.27?$$

But, I worked harder than needed

source: K. Roy, Approximate Computing: Ultra Low Power with "Good Enough" Results

Specifically, approximate Computing and Machine Learning form a perfect match since ML applications feature very high error resilience on one hand, while on the other they are able to fully exploit the hardware benefits derived from approximations achieving thus high area and power gains.

18 Bespoke Circuits

"Bespoke": fully customized circuit implementations

Low fabrication & NRE cost of PE ➡ Highly customizable architectures
ML classifiers *tailored* to the application

- Coefficients are hardwired in the circuit

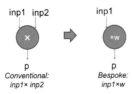

inp1 inp2   inp1

×   ×w

p   p
Conventional: Bespoke:
inp1× inp2   inp1×w

bespoke multiplier >5x lower area
[1]

Conventional	Bespoke
Inp1 [7:0] Inp2 [7:0]	Inp1 [7:0]

> | >C

Inp1 > Inp2 ? Inp1: Inp2   Inp1 > C ? Inp1: C

Area ~ 15 mm↑2   **Area ~ 3 mm↑2**
[2]

[1] G. Armeniakos, G. Zervakis, D. Soudris, M. Tahoori, J. Henkel, "Cross-Layer Approximation For Printed Machine Learning Circuits," DATE 2022
[2] K. Balaskas, G. Zervakis, K. Siozios, M. Tahoori, J. Henkel, "Approximate Decision Trees For Machine Learning Classification on Tiny Printed Circuits," ISQED 2022

Next, bespoke circuits refer to fully customized, per model implementations.
   The low fabrications costs in printed electronics combined with the on demand fabrication enable such a high customization that is mainly infeasible in silicon systems.

In such bespoke implementations the coefficients are hardwired in the circuit implementation itself leading to significant hardware gains.

For example, the area of a printed bespoke multiplier or comparator is 5 times smaller than the area of the respective conventional circuit.

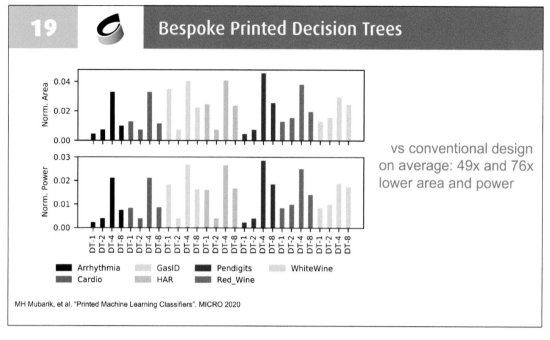

19 Bespoke Printed Decision Trees

vs conventional design
on average: 49x and 76x
lower area and power

MH Mubarik, et al. "Printed Machine Learning Classifiers". MICRO 2020

everaging the bespoke design paradigm, designing printed bespoke decision trees delivers high area and power gains that on average are as high as 49x and 76x respectively.

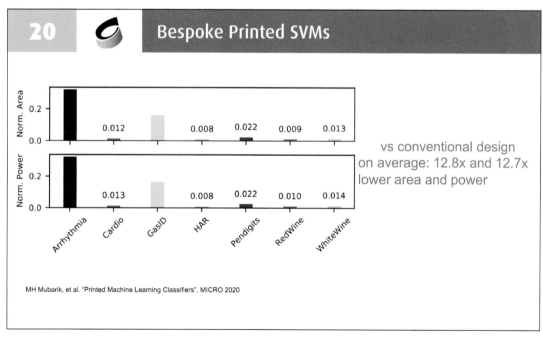

20 Bespoke Printed SVMs

vs conventional design
on average: 12.8x and 12.7x
lower area and power

MH Mubarik, et al. "Printed Machine Learning Classifiers". MICRO 2020

imilar results were obtained also for printed support vector machines.

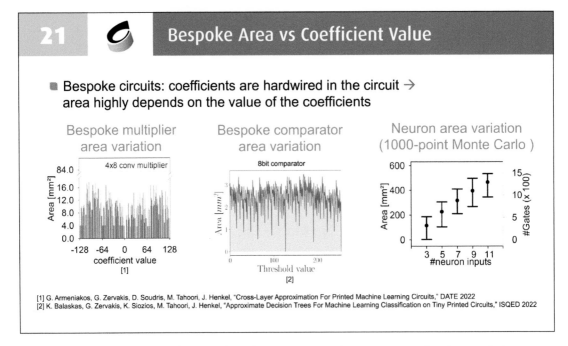

Integrating the coefficients in the circuit's implementation results in some peculiar characteristics.

Each circuit is unique and its associated hardware overheads are highly determined by the values of the coefficients.

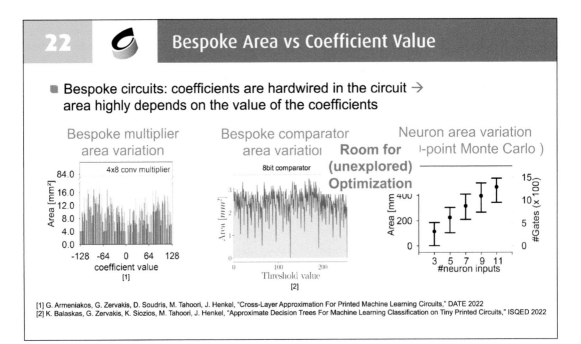

As a result, in printed bespoke circuits there is a lot of room for optimization.

As shown in these figures there is a high area variation with respect to the coefficient values. Thus, by selecting hardware-friendly coefficients we can achieve significant gains in area and consequently power consumption.

23 Approximate Printed MLP & SVM

- MLP and SVM: based on weighted sums

Approximations

- Precision scaling
  - weights (8bit) and inputs (4bit)
- Coefficient approximation
- Gate-level pruning

- Replace w with w', ∀ w: error & area are minimized
  - $w^+ \in [w-e, w]$ and $AREA(w^+)$ is min → positive error
  - $w^- \in [w, w+e]$ and $AREA(w^-)$ is min → negative error
  - replace w with $w' \in \{w^+, w^-\}$:
    $\Sigma(w - w')$ and $\Sigma(Area(w'))$ are minimized

FP model → low-precision FxP ⇨ Weight Approx. → HW-friendly weights ⇨ Bespoke Circuit ⇨ Synthesis ⇨ Gate-level Pruning

[1] G. Armeniakos, G. Zervakis, D. Soudris, M. Tahoori, J. Henkel, "Cross-Layer Approximation For Printed Machine Learning Circuits," DATE 2022

First, we examine printed multilayer perceptrons and support vector machines.

As approximation techniques we use precision scaling for the weights and inputs, coefficient approximation, and gate level pruning.

In gate-level pruning the circuit's gates that feature low switching activity and result in a bounded error by not being connected to the circuit's MSBs through any path, are replaced by a constant value.

In the coefficient approximation, the weights are replaced post-training with more hardware-efficient values.

Both the MLPs and SVMs are based on weighted sums. Hence, to implement the weight approximation we replace each weight with a larger or smaller one so that the area of the required multipliers minimizes and the positive errors due to the replacement balance the negative ones.

A brute force exploration is used to identify the optimal pruning and weight replacement configuration.

24 Approximate Printed Decision Trees

Approximations

- Precision scaling
  - thresholds and inputs (<8bit)
- Threshold approximation

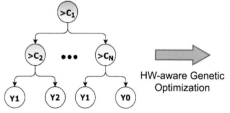

HW-aware Genetic Optimization

- Chromosome representation of each DT
- For each comparator C_i:
  - Selected bitwidth B_i
  - Selected range to look for threshold (±)

$2 * N$

| B_1 | ± | B_2 | ± | ... | B_N | ± |
| Comp 1 | | Comp 2 | | | Comp N | |

[1] K. Balaskas, G. Zervakis, K. Siozios, M. Tahoori, J. Henkel, "Approximate Decision Trees For Machine Learning Classification on Tiny Printed Circuits," ISQED 2022

Similarly, for the decision trees we use precision scaling and threshold approximation.

Though, in this case, each threshold may feature a different precision.

We use a genetic algorithm to find a close-to-optimal configuration for the precision of each comparator and approximate value for each threshold.

---

25 **Evaluation**

- ## Several Datasets from the UCI machine learning repository
  - http://archive.ics.uci.edu/ml/index.php

- ## MLPs, SVMs, DTs
  - ## Approximate Bespoke Designs vs Exact Bespoke ones

- ## Available open-source:
  https://github.com/garmeniakos/Ax-Printed-ML-Classifiers

For our evaluation we used the UCI machine learning repository and we examined SVMs, DTs and also MLPs.
Our work is distributed open-source in this repo.

---

26 **Evaluation Ax Printed MLPs & SVMs**

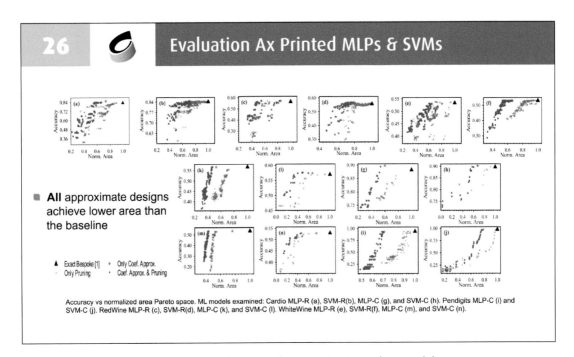

- **All** approximate designs achieve lower area than the baseline

▲ Exact Bespoke [1]　* Only Coef. Approx.
· Only Pruning　· Coef. Approx. & Pruning

Accuracy vs normalized area Pareto space. ML models examined: Cardio MLP-R (a), SVM-R(b), MLP-C (g), and SVM-C (h). Pendigits MLP-C (i) and SVM-C (j). RedWine MLP-R (c), SVM-R(d), MLP-C (k), and SVM-C (l). WhiteWine MLP-R (e), SVM-R(f), MLP-C (m), and SVM-C (n).

We evaluate our approximate printed MLPs and SVMs using in total 14 models.
As shown, the green points that apply both weight approximation and pruning are always on the pareto front and achieve always lower area than the baseline.

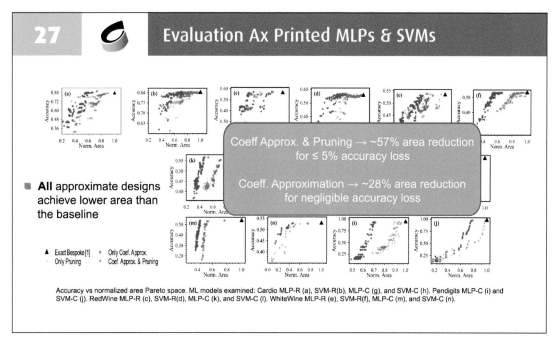

27 — **Evaluation Ax Printed MLPs & SVMs**

- **All** approximate designs achieve lower area than the baseline

> Coeff Approx. & Pruning → ~57% area reduction for ≤ 5% accuracy loss
>
> Coeff. Approximation → ~28% area reduction for negligible accuracy loss

▲ Exact Bespoke [1]   ∗ Only Coef. Approx.
· Only Pruning   ∗ Coef. Approx. & Pruning

Accuracy vs normalized area Pareto space. ML models examined: Cardio MLP-R (a), SVM-R(b), MLP-C (g), and SVM-C (h). Pendigits MLP-C (i) and SVM-C (j). RedWine MLP-R (c), SVM-R(d), MLP-C (k), and SVM-C (l). WhiteWine MLP-R (e), SVM-R(f), MLP-C (m), and SVM-C (n).

S pecifically, our approximate SVMs and MLPs achieve 57% area reduction for less than 5% accuracy loss. We must note that applying only our proposed coefficient approximation delivers 28% area reduction for almost zero accuracy loss.

---

28 — **Evaluation Ax Printed MLPs & SVMs**

- AREA AND POWER EVALUATION FOR LESS THAN 1% ACCURACY LOSS. HIGHLIGHTED DESIGNS CAN BE POWERED BY A MOLEX 30MW BATTERY

ML Circuit	Coeff. Approx. & Pruning			Only Coeff. Approx.			Only Pruning		
	A^1	P^2	$AG^3 PG^3$	A^1	P^2	$AG^3 PG^3$	A^1	P^2	$AG^3 PG^3$
Card MLP-R	12	37	45 44	16	49	27 26	18	56	16 15
Card SVM-R	3.5	13	49 42	5.5	19	19 15	5.0	18	26 22
RW MLP-R	3.3	12	53 49	6.0	21	15 14	4.6	17	35 30
RW SVM-R	2.6	10	35 33	3.1	12	22 22	2.9	11	27 25
WW MLP-R	8.0	27	39 35	11	34	20 17	9.2	29	30 28
WW SVM-R	2.2	8.5	47 45	2.8	11	34 32	3.4	13	19 19
Card MLP-C	17	54	48 44	20	62	40 36	33	97	0 0
Card SVM-C	8.7	29	43 38	10	33	33 29	14	43	8.7 8.3
Pend MLP-C	46	153	31 28	48	143	29 33	60	194	10 9.0
Pend SVM-C	97	287	22 21	97	287	22 21	121	357	2.2 1.8
RW MLP-C	8.0	27	55 50	9.3	30	47 43	18	53	0 0
RW SVM-C	7.6	26	68 65	16	50	32 31	15	49	35 33
WW MLP-C	13	42	57 57	24	73	23 26	16	52	47 48
WW SVM-C	11	36	61 59	21	65	26 25	15	46	49 47

[1] Area (cm²). [2] Power (mW). [3] Area and Power Gain compared to the bespoke baseline [1] (in %).

- BASELINE CIRCUITS

	MLP-C					MLP-R					SVM-C					SVM-R				
	Acc[1]	T[2]	#C[3]	Area (cm²)	Power (mW)	Acc[1]	T[2]	#C[3]	Area (cm²)	Power (mW)	Acc[1]	T[2]	#C[3]	Area (cm²)	Power (mW)	Acc[1]	T[2]	#C[3]	Area (cm²)	Power (mW)
Cardio	0.88	(21,3,3)	72	33.4	97.3	0.83	(21,3,1)	66	21.6	65.9	0.90	3	63	15.1	46.8	0.84	1	21	6.8	22.9
Pendigits	0.94	(16,5,10)	130	67.0	213.0	0.37	(16,5,1)	85	.[4]	-	0.98	45	160	123.8	364.4	0.23	1	16	.[4]	-
RedWine	0.56	(11,2,6)	34	17.6	55.3	0.56	(11,2,1)	24	7.1	24.0	0.57	15	66	23.5	72.9	0.56	1	11	4.0	15.1
WhiteWine	0.54	(11,4,7)	72	31.2	98.4	0.53	(11,4,1)	48	13.1	40.7	0.53	21	77	28.3	87.4	0.53	1	11	4.2	15.5

[1] Accuracy using 8-bit coefficients and 4-bit inputs. [2] Model's topology (for SVMs: the number of classifiers). [3] Number of coefficients of the model.
[4] These models achieve low accuracy and are not evaluated.

- 47% area & 44% power reduction for ≤ 1% acc loss

- Standalone Coeff.Approx. → 28% & 26%
- Standalone Pruning → 22% & 20%

- Highlighted (green) designs can now be powered by a printed battery (<30mW)

I n this table we consider a 1% accuracy loss threshold and we highlight in green the printed classifiers that can be battery powered.

A Molex 30mW battery is used.

As shown, almost the 60% of our approximate classifiers can be battery powered.

On the other hand, only the 30% of the exact classifiers could be power by an existing printed battery.

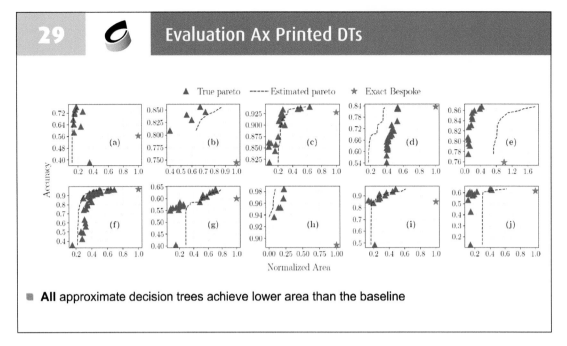

As expected, we obtained similar results also for the approximate decision trees. Again all approximate decision trees achieve lower area than the baseline.

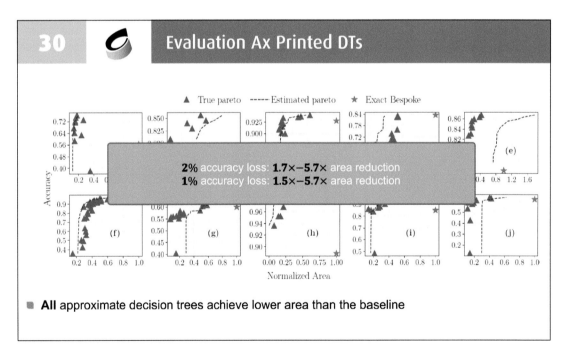

Specifically, up to 5.7x lower area can be achieved for les than 2% accuracy loss.

31 Evaluation Ax Printed DTs

■ Evaluated solutions for 1% accuracy loss

■ Power **and** area efficiency:
  ■ Average 3.2× area reduction
  ■ Average 3.4× power reduction

Dataset	Accuracy	Area (mm^2)	Norm. Area	Power (mW)	Norm. Power
Arrhythmia	0.67	22.30	0.137	1.04	0.138
Balance	0.81	27.28	0.401	1.16	0.372
Cardio	0.92	43.54	0.244	2.05	0.253
HAR	0.83	294.54	0.534	13.70	0.525
Mammogr.	0.81	8.06	0.082	0.38	0.084
PenDigits	0.96	368.48	0.641	16.10	0.644
Redwine	0.60	267.21	0.520	11.70	0.525
Seeds	0.94	2.32	0.077	0.09	0.064
Vertebral	0.86	7.84	0.136	0.38	0.142
WhiteWine	0.61	124.11	0.229	5.35	0.230

■ Powered by Blue Spark printed batteries ($<3mW$) → green

■ Self powered / harvesters ($<0.1mW$) → orange

Considering the 1% accuracy loss threshold we evaluate the battery powered operation of the printed decision trees.

As shown in this table the 50% of the examined decision tree classifiers can be powered by only a 3mW blue spark battery while the Seeds decision tree can be powered by a printed harvester.

32 Conclusion

■ PE offer a solution to application domains with limited infiltration of computing

■ Designing Printed ML circuits is extremely challenging due to the inherent limitations
  ■ mandates alternative computing methods and high customization

■ Codesign Approximate ML circuits shows promising results towards the realization of Printed ML circuits

Concluding, printed electronics establish as a solution to enable smart services in application domains that haven't witnessed computing infiltration up to now.

Though, printed electronics form an extreme use case of embedded machine learning and due to several limitations the design of complex printed classifiers is very challenging and mandates non conventional approaches.

As we demonstrated, a software hardware codesign that exploits fully customized circuit implementations and applies approximations in both levels delivers promising results towards the realization of printed ML classification.

Still, there is a lot of room for further improvement.

CHAPTER 05

The Analog Designer's Toolbox (ADT) Towards a New Paradigm for Analog IC Design

Prof. Hesham Omran

Ain Shams University and Master Micro LLC, Egypt

1 **Outline**

- [ ] **Introduction**
- [ ] **The gm/ID Design Methodology**
- [ ] **What Is ADT?**
- [ ] **Why ADT?**

This is the outline of the presentation. We will start with a general introduction followed by an overview about the GM/ID design methodology. We will then discuss ADT in detail.

2 **Introduction**

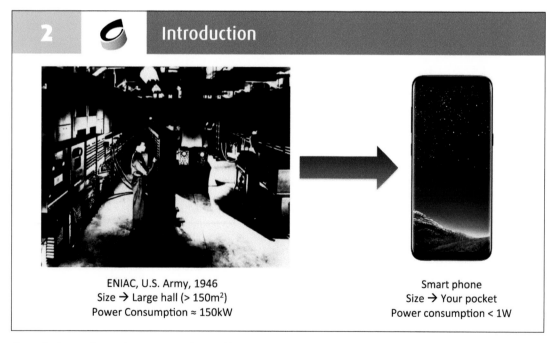

ENIAC, U.S. Army, 1946
Size → Large hall (> 150m²)
Power Consumption ≈ 150kW

Smart phone
Size → Your pocket
Power consumption < 1W

But before talking about ADT, I always like to start by showing this picture of ENIAC, the first general purpose electronic computer. And the reason I like to show this picture is that when you compare it with the smart phone that everyone now carries in his pocket, you feel how much tremendous change has happened in the field of electronics in the last 70 years.

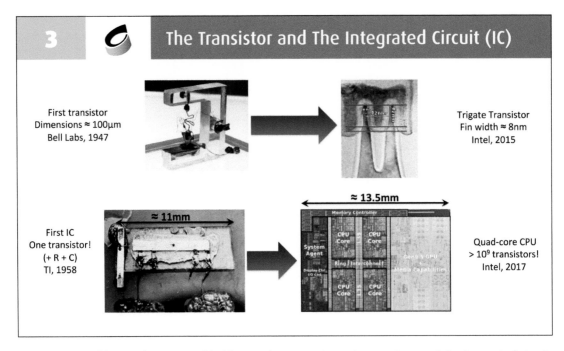

And as we all know, this was enabled by two key inventions, the transistor and the integrated circuit, and each on of these two inventions also witnessed exponential advancement in the last decades.

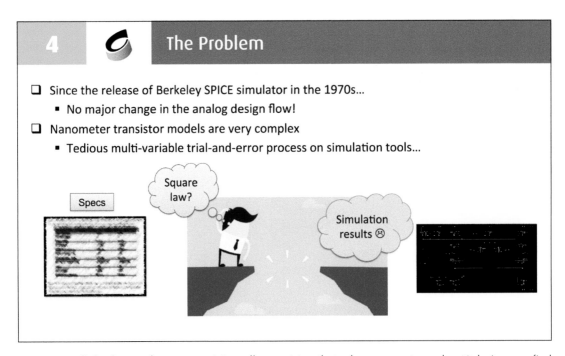

So given all this huge advancement it is really surprising that when we go to analog IC design, we find that the analog design flow has almost been the same for the last 50 years.

And the problem is getting worse because the nanometer transistor models are very complex, so the design process is basically becoming a trial and error process on the simulation tools and there is no systematic methodology for doing analog design.

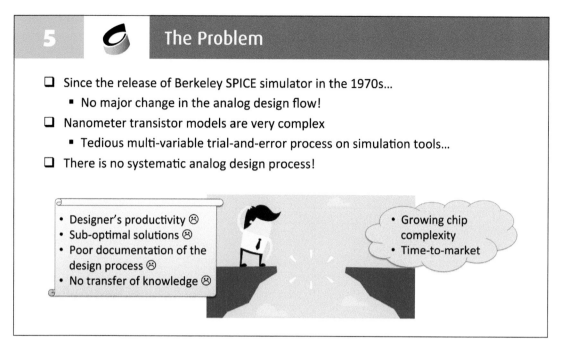

5   **The Problem**

❑ Since the release of Berkeley SPICE simulator in the 1970s...
  ▪ No major change in the analog design flow!
❑ Nanometer transistor models are very complex
  ▪ Tedious multi-variable trial-and-error process on simulation tools...
❑ There is no systematic analog design process!

- Designer's productivity ☹
- Sub-optimal solutions ☹
- Poor documentation of the design process ☹
- No transfer of knowledge ☹

- Growing chip complexity
- Time-to-market

And this creates a large gap between from one side the growing chip complexity and the tight time to market requirements and from the other side the designers have limited productivity, so they just try to get any working solution for their circuit even if it is a sub-optimal solution that is wasting lots of power or area. And there is no documentation for the thinking process of the designer, and this makes it very difficult to transfer the knowledge and expertise from the senior generations of designers to the junior generations of designers. And that's why students and junior designers see analog design as black magic, they don't really know what is going on and it is very difficult for them to start a design from scratch.

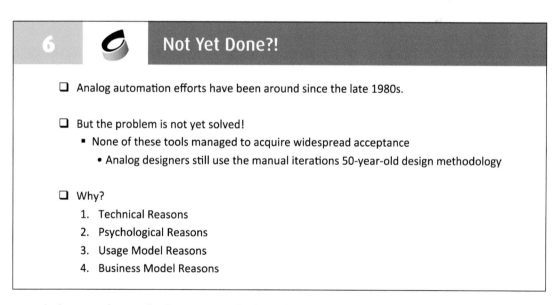

6   **Not Yet Done?!**

❑ Analog automation efforts have been around since the late 1980s.

❑ But the problem is not yet solved!
  ▪ None of these tools managed to acquire widespread acceptance
    • Analog designers still use the manual iterations 50-year-old design methodology

❑ Why?
  1. Technical Reasons
  2. Psychological Reasons
  3. Usage Model Reasons
  4. Business Model Reasons

And of course, I'm not the first one to talk about this problem. People have been trying to automate analog design for more than 30 years. But we can say that the problem is not yet solved, and the evidence is that the majority of the analog design community is still doing manual iterations on SPICE simulation tools, the same 50 year old design methodology.

So, we need some reflection: Why none of the solutions offered by the academia and the industry managed to solve the problem. And we categorized the reasons into four categories: technical, psychological, usage model, and business model.

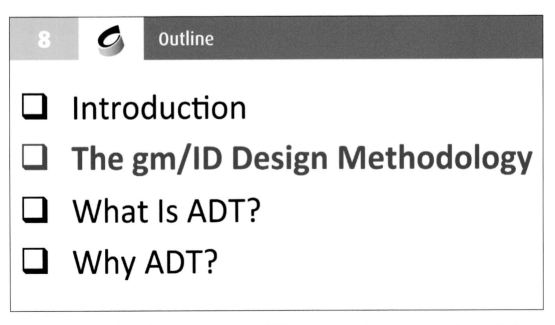

7 — The Solution: The Analog Designer's Toolbox (ADT)

Technical	Psychological	Usage Model	Business Model
• Precomputed look-up tables technology • Custom vectorized solvers • Very fast execution time • Simulator-accurate results	• We are not replacing the designers • Make the designers more productive. • Make the design process systematic, optimized, and fun!	• From Designers... To Designers • User-friendly designer-oriented interface • Short learning curve • Demos and quick-start videos	• Academia and industry • Free license for academia (the next gen of designers)

And these four categories are what we are addressing with our new solution: the analog designers toolbox ADT. From the technical side we are using the precomputed lookup tables and custom vectorized solvers, and this gives us two unique advantages: very fast execution time and simulator accuracy. From the psychological side we are not trying to replace the designers. On the contrary we are empowering the designers to make them more productive and to make the whole process systematic, optimized, and fun. From the usage model side, our tool is developed by designers to designers, so we have a very friendly user interface and a short learning curve. From the business model side, we are targeting both academia and the industry, and we are offering a free license for the next generations of designers to invite everyone to join us in this new paradigm.

8 — Outline

- ☐ Introduction
- ☐ **The gm/ID Design Methodology**
- ☐ What Is ADT?
- ☐ Why ADT?

But before talking about the details of ADT, I would like to give a brief overview about the gm/ID design methodology which is one of the most popular analog design methodologies and it is seamlessly built in inside out tool.

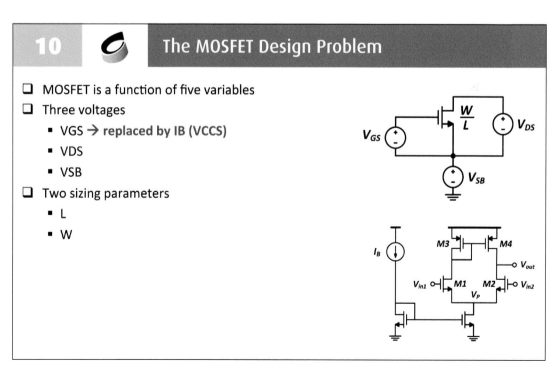

9 The MOSFET Design Problem

❑ MOSFET is a function of five variables
❑ Three voltages
- VGS
- VDS
- VSB
❑ Two sizing parameters
- L
- W

If we think about the MOSFET as a design problem, it is basically a function of five variables. These five variables are the three terminal voltages: VGS, VDS, VSB, and the two sizing parameters: the channel length and the channel width. And when you are designing an analog circuit, you actually need to specify these five variables for every transistor in the circuit you are designing.

10 The MOSFET Design Problem

❑ MOSFET is a function of five variables
❑ Three voltages
- VGS → replaced by IB (VCCS)
- VDS
- VSB
❑ Two sizing parameters
- L
- W

So let's see how does this happen. First we don't specify the VGS directly because most of the time we bias our circuits using current sources and current mirrors like the one shown in the figure. So the VGS degree of freedom is replaced by the bias current degree of freedom because the transistor is basically a voltage controlled current source.

11   The MOSFET Design Problem

- ❏ MOSFET is a function of five variables
- ❏ Three voltages
  - VGS → **replaced by IB (VCCS)**
  - VDS → Set to VDsat + margin (limited headroom)
  - VSB → Usually imposed by circuit topology
- ❏ Two sizing parameters
  - L
  - W

VDS and VSB both have secondary importance because we just want to make sure the device is biased in saturation by some margin, and VSB, most of the time, is imposed by the circuit topology. So the designer is left with the two sizing parameters L and W.

12   Selecting L: Use Your Designer's Intuition!

Use shorter L if you want	Use longer L if you want

➢ Smaller area
 ➢ Smaller capacitance
 ➢ High speed (high $f_T = \frac{g_m}{2\pi C_{gg}}$)
	➢ High r_o (high V_A)
	     ▪ Must have large VDsat margin to be effective (beware of exceptions due to feedback)
	 ➢ Less random mismatch
	     ▪ Longer L implies larger area (beware of exceptions due to non-uniform doping profile)
	 ➢ Low flicker noise
	     ▪ Longer L implies larger area

When's selecting L you can use your designer's intuition based on the trade-offs you are trying to address. So, if you care more about area, capacitance, and speed you will pick a short L. And if you care more about output resistance, mismatch, and flicker noise, you will pick a long L.

13 | **Selecting W: A Nonintuitive Variable** ☹

- ❏ Choosing W is one of the most difficult tasks
- ❏ Assume you select W = 100 um
  - ▪ Is this large or small?!
- ❏ The answer depends on a bunch of parameters!
  - ▪ What L do you use?
  - ▪ What ID do you force?
  - ▪ What gm/ID do you want (inversion level: WI, MI, SI)?
  - ▪ What technology node, the device flavor, etc.
- ❏ The search-range of W can range from sub-1um to 1000um (laid-out as multi-fingers)

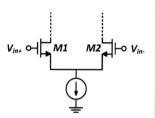

But the big problem is when selecting W because it is an unintuitive variable. And to give an example, if I tell you that I'm using 100 um widths for this diff pair, do you consider this a large value or a small value? Actually there is no clear answer to this question because the answer depends on a bunch of parameters. It depends on what is your channel length? What is your bias current? what is your inversion level? Whether you want to bias your device in weak inversion or moderate inversion or strong inversion? And it also depends on the technology node, the device flavor. And what makes the problem worse is that the search range for the width can be more than three orders of magnitude. So it is very difficult to make an educated intuitive initial guess for the device width.

 14 | **The Old Fix: Vov**

- ❏ To solve this problem, designers used to replace W by Vov in the DOFs
- ❏ Vov = VGS – VTH used to be related to circuit specs (gain, speed, noise, headroom, etc.)
- ❏ Given ID, L, and Vov: Use the square law to calculate W

$$I_D = \frac{\mu_n C_{ox}}{2} \frac{W}{L} V_{ov}^2$$

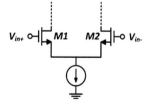

- ❏ This fix doesn't work any more
  - ▪ The square law is not accurate in SI
    - • And completely invalid in MI and WI
  - ▪ Vov is not related to circuit specs anymore: $g_m \neq \frac{2I_D}{V_{ov}}$ and $V_{Dsat} \neq V_{ov}$
    - • No direct relation to gain, speed, noise, headroom, etc.

And the old fix for this problem was using the concept of the overdrive voltage. The designers used to specify Vov instead of specifying the width, and then they plugged in numbers in the famous and simple square law equation to get a rough estimate for the width.

But as we all know, this fix this old fix doesn't work anymore because the square low is not accurate for short channel devices, and it is completely meaningless for both long and short devices if you bias your device in moderate or weak inversion. So the overdrive voltage concept is no more related to the circuit parameters we care about as designers.

15 The Modern Fix: The gm/ID Design Methodology

- ❑ The gm/ID captures the relation between:
    - The fundamental function of the transistor: the transconductance (gm)
    - The most valuable resource: the power consumption (ID)

> **The Transistor Efficiency**
>
> $$TE = \frac{What\ you\ want}{What\ you\ pay} = \frac{g_m}{I_D}$$

- ❑ The gm/ID is directly related to the most important analog specs
    - Speed, noise, efficiency, gain, swing, mismatch!
- ❑ What we care most about is the gm/ID!
- ❑ Replace W by gm/ID in the DOFs
- ❑ **Think gm/ID!**

And the modern fix to this problem is the GM/ID design methodology. The GM/ID captures the relation between the most important property of the transistor, the GM, the transconductance, and the most valuable resource, the power consumption. So if we think about the transistor as a system, the GM is the output that we want, and the ID is the price that we pay to get this output. And this GM/ID ratio is directly related to the circuit specs we care about as designers, so this is the real design knob that we should use to size our transistors.

16 Think gm/ID!

- ❑ The gm/ID is intuitive with a limited search-range
    - Typically: 5 to 25
    - The range of gm/ID values doesn't differ much
        - From one device to another
        - And from one technology to another

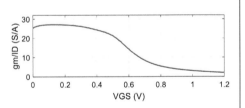

- ❑ The gm/ID is an "Orthogonal" and normalized control of TE!
    - And consequently, the inversion level: WI, MI, SI
- ❑ When ID and L are changed
    - gm/ID (the TE) is kept unchanged
    - We simply lookup the new W

And the good news is that the search range for the GM/ID is very limited, typically from 5 to 25. And this range doesn't differ from one device to another or from one technology to another. So we can think of the GM/ID as an orthogonal and normalized parameter to control the transistor efficiency or equivalently to define the transistor inversion level. And the GM/ID is orthogonal because if we change the ID or the L, as long as we keep the GM/ID fixed, the bias point is actually fixed, and we simply look up the new W that corresponds to this change.

17   Think gm/ID: Designer's Intuition Restored!

Use small gm/ID if you want	Use large gm/ID if you want

➢ Strong-inversion (SI) biasing
➢ Small gm (for a given ID)
   • Devices whose gm does NOT contribute to gain (Ex: active loads)
➢ Small area
➢ Small capacitance
➢ High speed
➢ High r_o (high V_A)
   • VDS has less effect on current
	➢ Moderate inversion (MI) or weak-inversion (WI) biasing
	➢ Large gm (for a given ID)
	   • Devices whose gm contributes to gain (Ex: input stage and cascode devices)
	➢ High efficiency
	   • Low power consumption (low ID) for a given speed or noise spec (gm spec)
	➢ Less random mismatch
	   • Large gm/ID implies larger W (larger area) (beware of systematic mismatch)
	➢ Low flicker noise
	   • Large gm/ID implies larger W (larger area)
	➢ Large input range and/or output swing
	   • Large gm/ID implies small VDsat $\approx V^* = \frac{2}{g_m/I_D}$

So the GM/ID is all about restoring the designer's intuition, because if you care more about area, capacitance, and speed you will pick a small GM/ID and go towards strong inversion. And if you care more about efficiency, mismatch, and flicker noise you will pick a large GM/ID and go towards moderate or weak inversion. Most of the time, the best compromise is in using MI.

18   The MOSFET DOFs

Original	The Old-School	The gm/ID Methodology
W	Vov (square law)	gm/ID (use charts or LUTs to get W)
L	L (trial and error)	L (use charts or LUTs)
VGS	ID (current mirror biasing)	ID (current mirror biasing)
VDS	VDS = Vov + VDsat_margin (trial and error)	VDS = VDsat + VDsat_margin (taken into account by using charts or LUTs)
VSB	Forced by topology (use simple model or ignore)	Forced by topology (taken into account by using charts or LUTs)

To summarize this point the old school used to replace the W with the overdrive voltage, and then they used the square law and some trial and error to get the other parameters. The GM/ID is all about replacing the non intuitive width variable and the expired overdrive voltage concept with the GM/ID ratio, and then we can use design charts or lookup tables to get the width and all other parameters while preserving the simulator accuracy.

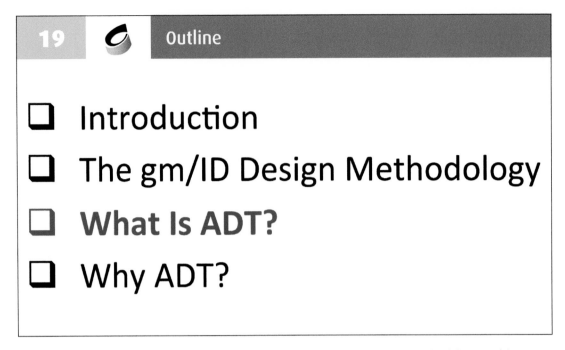

Now we are ready to talk about how ADT integrates the GM/ID design methodology and how it is disrupting analog IC design.

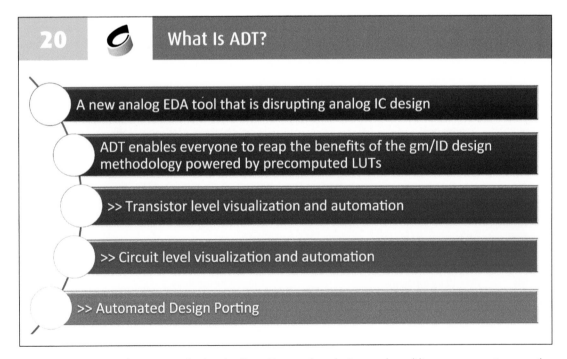

ADT is a new analog EDA tools that is disrupting analog design and enabling everyone to reap the benefits of the GM/ID design methodology with minimal effort. ADT is addressing three problems: transistor level design, circuit level design, and automated design porting.

21 — What Is ADT?

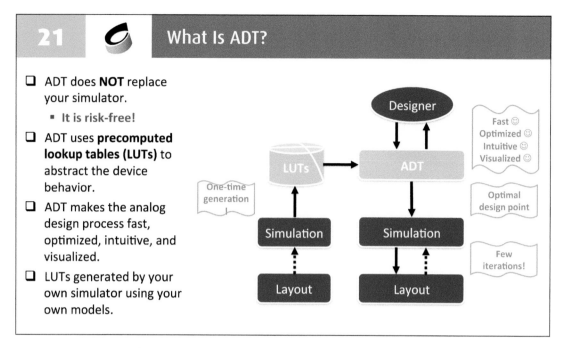

- ❑ ADT does **NOT** replace your simulator.
  - ▪ It is risk-free!
- ❑ ADT uses **precomputed lookup tables (LUTs)** to abstract the device behavior.
- ❑ ADT makes the analog design process fast, optimized, intuitive, and visualized.
- ❑ LUTs generated by your own simulator using your own models.

t is important to note that ADT does not replace the conventional simulation tools. So it doesn't add any risk to the design process. ADT adds a new layer between the designer and the simulation tools and this new layer is powered by precomputed lookup tables. The goal of this layer is to avoid that lengthy iterations in the design cycle by providing the designer with an optimal design point in a fast, intuitive, and visualized way.

22 — Outline

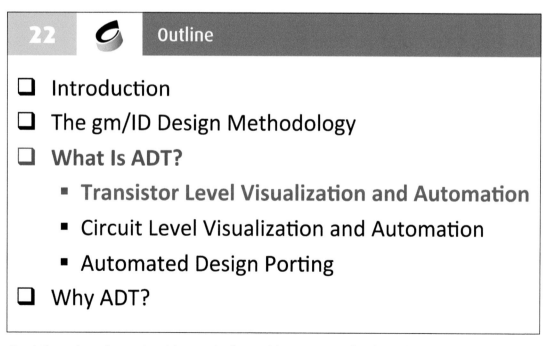

- ❑ Introduction
- ❑ The gm/ID Design Methodology
- ❑ What Is ADT?
  - ▪ **Transistor Level Visualization and Automation**
  - ▪ Circuit Level Visualization and Automation
  - ▪ Automated Design Porting
- ❑ Why ADT?

So let's see how those ADT addresses the first problem: transistor level visualization and automation.

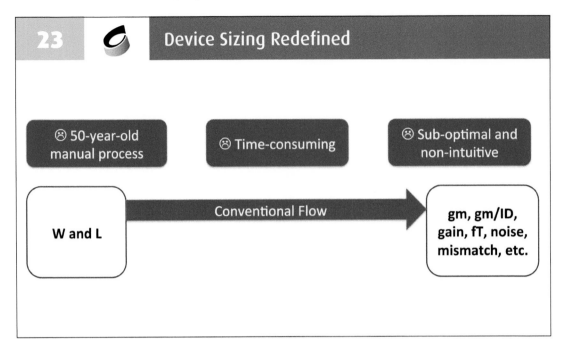

The conventional way of doing device sizing is that the designer has control on the device geometry (W and L). But actually he cares about the device performance parameters like the GM, the gain, the FT, the noise. So he has to tweak what he doesn't know in a manual time-consuming process in order to get what he really cares about.

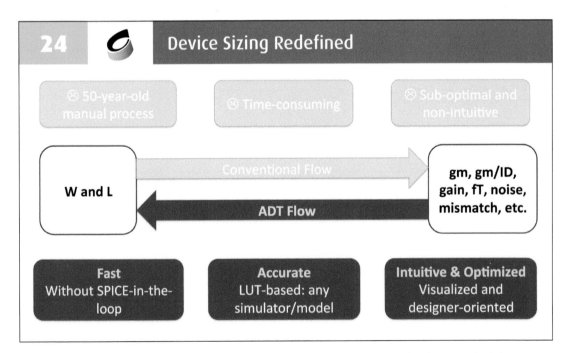

And ADT reverses this flow so that it makes sense! Instead of starting with what you don't know, ADT enables you to start with what you know, what you really care about, and then it tells you about the device geometry which is what you don't know, in very fast, accurate, intuitive, and optimized way.

---

25   **Outline**

- ❏ Introduction
- ❏ The gm/ID Design Methodology
- ❏ **What Is ADT?**
  - ▪ **Transistor Level Visualization and Automation**
    - • **ADT Device Xplore**
    - • **ADT Sizing Assistant**
  - ▪ **Circuit Level Visualization and Automation**
  - ▪ **Automated Design Porting**
- ❏ **Why ADT?**

And to achieve this target we have two tools. The first one is ADT device explorer and the second one is ADT sizing assistant.

---

26   **Design Example: IGS**

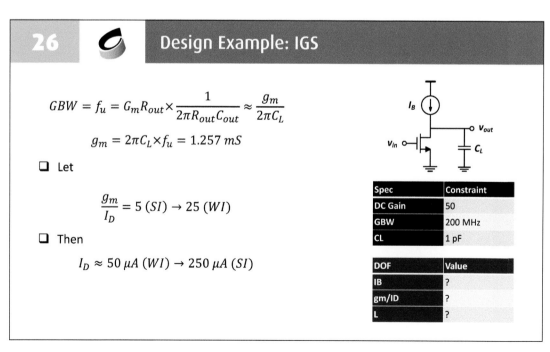

$$GBW = f_u = G_m R_{out} \times \frac{1}{2\pi R_{out} C_{out}} \approx \frac{g_m}{2\pi C_L}$$

$$g_m = 2\pi C_L \times f_u = 1.257 \ mS$$

- ❏ Let

$$\frac{g_m}{I_D} = 5 \ (SI) \rightarrow 25 \ (WI)$$

- ❏ Then

$$I_D \approx 50 \ \mu A \ (WI) \rightarrow 250 \ \mu A \ (SI)$$

Spec	Constraint
DC Gain	50
GBW	200 MHz
CL	1 pF

DOF	Value
IB	?
gm/ID	?
L	?

Let's take a design example to illustrate this. This is a simple intrinsic game stage and assume we have spec on the gain, the GBW, the gain bandwidth product (GBW) and the load capacitance. For this single transistor amplifier we have 3 degrees of freedom: the bias current, the GM/ID and the L. Given the GBW spec we can compute the GM, and because the GM/ID has a limited search range, we can get a quick estimate of how much bias current do we need.

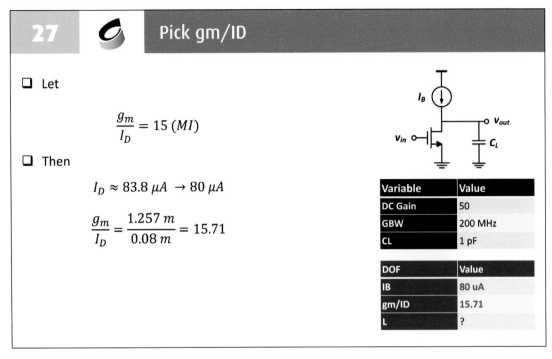

27 **Pick gm/ID**

- ❑ Let

$$\frac{g_m}{I_D} = 15 \; (MI)$$

- ❑ Then

$$I_D \approx 83.8 \; \mu A \; \rightarrow 80 \; \mu A$$

$$\frac{g_m}{I_D} = \frac{1.257 \; m}{0.08 \; m} = 15.71$$

Variable	Value
DC Gain	50
GBW	200 MHz
CL	1 pF

DOF	Value
IB	80 uA
gm/ID	15.71
L	?

I f we pick a GM/ID in moderate inversion, we can compute the bias current, round it to a nice number, and then recompute the GM/ID.

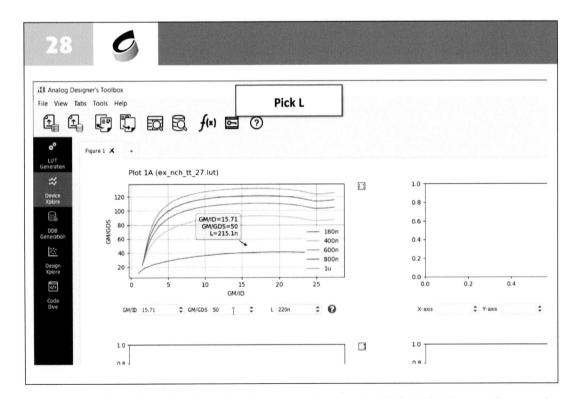

28 **Pick L**

N ow we need to choose the device length. We can visualize the GM/GDS (intrinsic gain) versus the GM/ID and find the L that corresponds to the intrinsic gain we are looking for using the magic cursor that can move between traces.

29

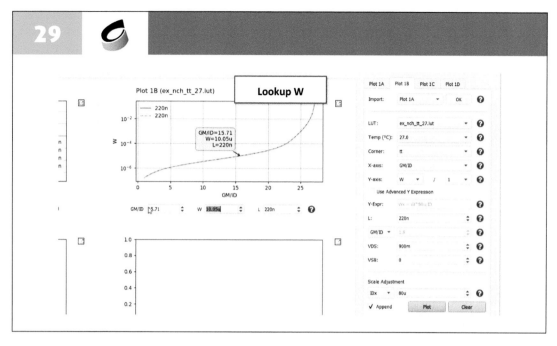

We can also plot the W vs the GM/ID and find the W that corresponds to the GM/ID and L we have selected. This way we can find W across all operating regions of the transistor using simulator accurate data.

30   Testbench and Results

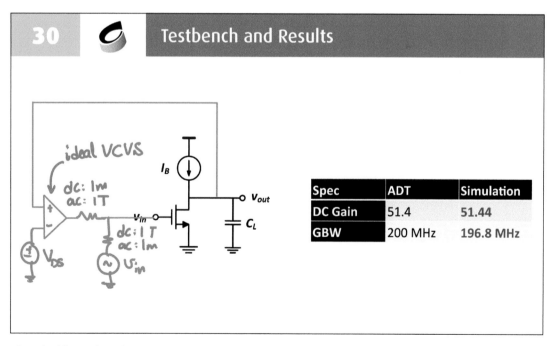

Spec	ADT	Simulation
DC Gain	51.4	51.44
GBW	200 MHz	196.8 MHz

If we build a testbench on SPICE, we must use a feedback loop as a simulation trick to set the bias voltage at the high impedance output node. The DC Gain from ADT matches the simulation. The GBW is slightly degraded because we ignored the device parasitic capacitance.

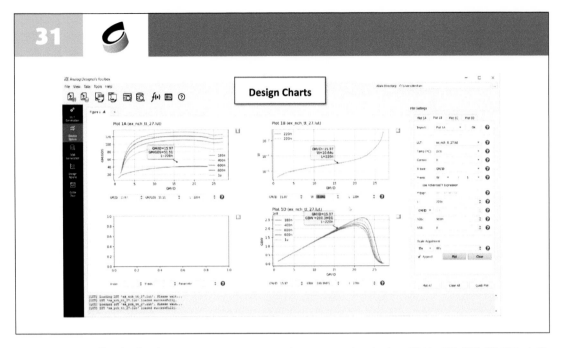

We can actually take the device parasitic capacitance into account by plotting GBW = GM/(2*pi*(CDD+1p)). The GBW increases as we increase the GM/ID, peaks at some point that is L dependent, then starts degrading as we go into deep subthreshold where the parasitics increase exponentially.

32 Final Design

DOF	Value
IB	80 uA
gm/ID	15.97
L	0.22 um

Spec	ADT	Simulation
DC Gain	51.5	51.53
GBW	200 MHz	199.8 MHz

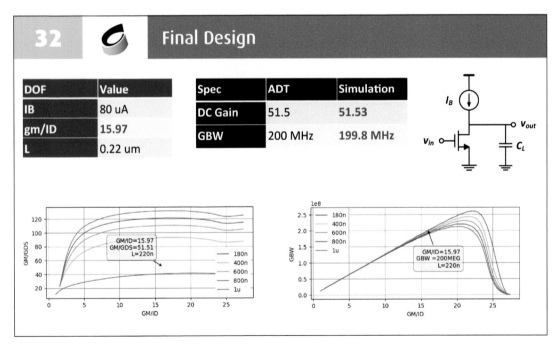

Now we have the final design, and the simulation results almost exactly match ADT results. This is not a surprise because ADT uses the LUTs that are generated by the simulator!

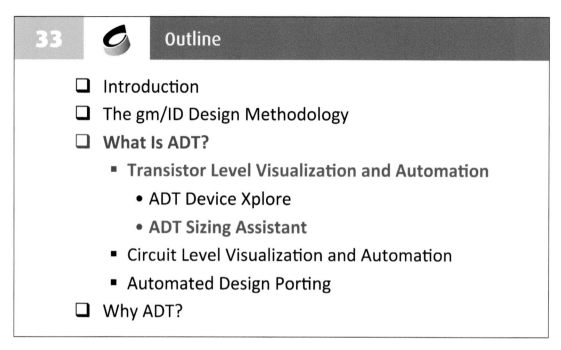

33 Outline

- ❏ Introduction
- ❏ The gm/ID Design Methodology
- ❏ **What Is ADT?**
  - ▪ **Transistor Level Visualization and Automation**
    - • ADT Device Xplore
    - • **ADT Sizing Assistant**
  - ▪ Circuit Level Visualization and Automation
  - ▪ Automated Design Porting
- ❏ Why ADT?

The second solution to automate sizing at the transistor level is using the ADT sizing assistant.

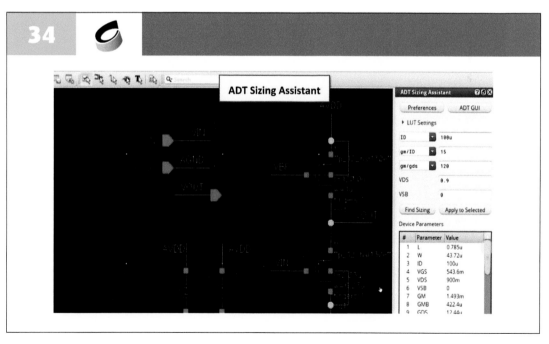

34

ADT sizing assistant is integrated inside the schematic editor to enable the designer to find the sizing very quickly. The designer can define the parameters that he wants the device to satisfy. The assistant will find the sizing and all the operating parameters of the device. The sizing can then be applied directly to the selected device in the schematic.

35 Outline

- ❑ Introduction
- ❑ The gm/ID Design Methodology
- ❑ **What Is ADT?**
  - ▪ **Transistor Level Visualization and Automation**
  - ▪ **Circuit Level Visualization and Automation**
  - ▪ **Automated Design Porting**
- ❑ Why ADT?

A DT also addresses circuit level sizing by novel visualization and automation.

36 Circuit Level: Conventional Flow

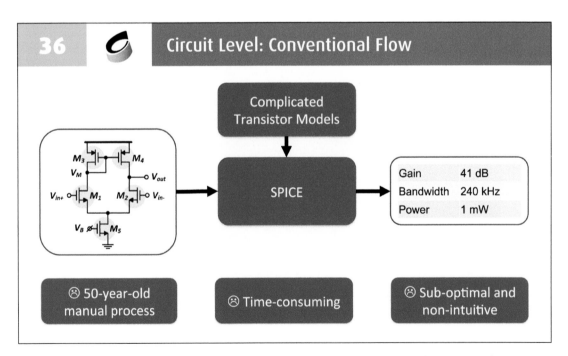

The conventional way of doing it is manually tweaking the W and L of individual transistors, then running SPICE simulations looking for the circuit specs in a manual, time-consuming, and sub-optimal process.

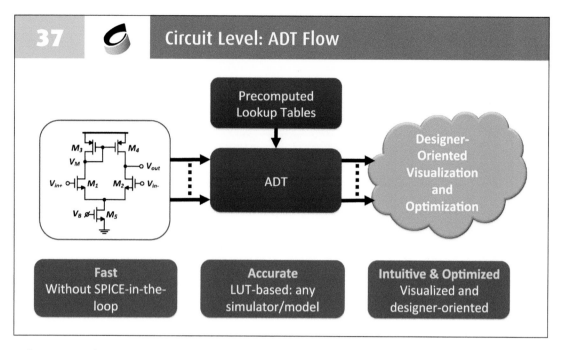

The way it is done in ADT is that the circuit is solved hundreds of thousands of times to paint the design space, and enable the designer to visualize the design trade-offs and select the global optimal design point in a fast and accurate way.

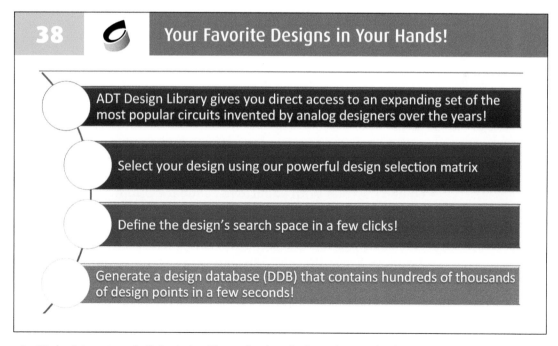

ADT also integrates a built-in design library that has the best designs that have been invented by analog designers over the years. So instead of being lost in text books and research papers, you can quickly pick your design and generate a design database (DDB) that has hundreds of thousands of design points.

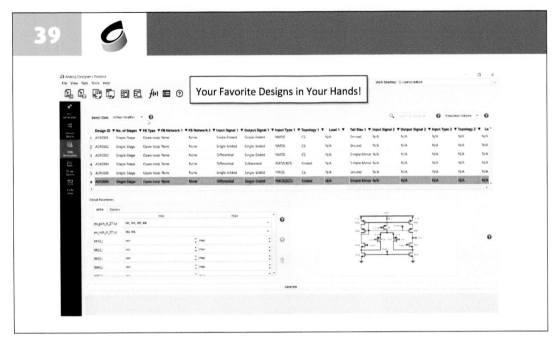

The built in library has amplifiers, voltage regulators, reference circuits, and current mirrors. The analog building blocks we commonly use. And you can select the properties of your design using filters similar to an Excel sheet! The search range of every transistor is defined by L and GM/ID, not L and W.

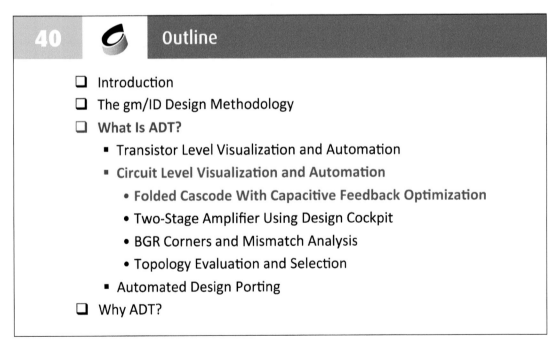

Outline

- ❑ Introduction
- ❑ The gm/ID Design Methodology
- ❑ What Is ADT?
  - ▪ Transistor Level Visualization and Automation
  - ▪ **Circuit Level Visualization and Automation**
    - • **Folded Cascode With Capacitive Feedback Optimization**
    - • Two-Stage Amplifier Using Design Cockpit
    - • BGR Corners and Mismatch Analysis
    - • Topology Evaluation and Selection
  - ▪ Automated Design Porting
- ❑ Why ADT?

Let's take some practical design examples to show much powerful ADT can be.

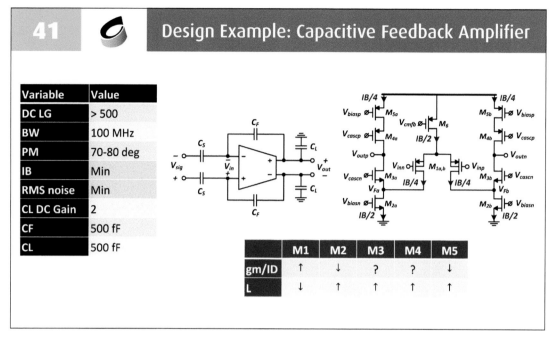

Design Example: Capacitive Feedback Amplifier

41

Variable	Value
DC LG	> 500
BW	100 MHz
PM	70-80 deg
IB	Min
RMS noise	Min
CL DC Gain	2
CF	500 fF
CL	500 fF

	M1	M2	M3	M4	M5
gm/ID	↑	↓	?	?	↓
L	↓	↑	↑	↑	↑

This is a capacitive feedback amplifier that uses a folded cascode topology. A very popular circuit in data converters and filters. We have several constraints and two objectives: power consumption and noise. The expert designer can give some qualitative directions, but it is very hard to give quantitative answers.

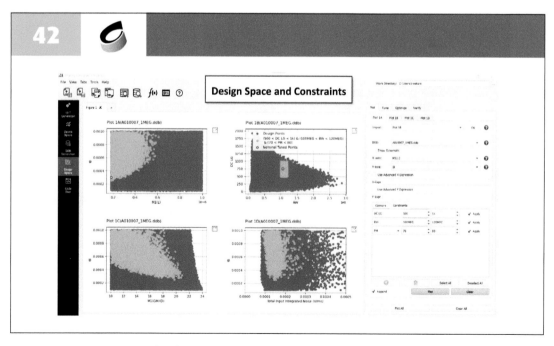

42

Design Space and Constraints

Using ADT you can paint the design space and give quantitative answers in a matter of seconds. The orange points are the points satisfying the constraints. We can quickly see what is the minimum power consumption, and how the power trades-off with the noise.

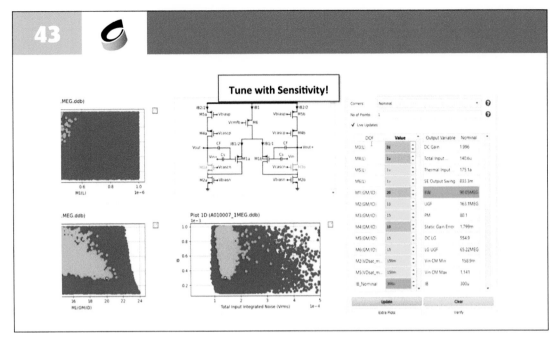

We cab also run interactive sensitivity analysis so we see what are the parameters affecting each spec, and in which direction should we tune each parameter. Some decisions may counterintuitive like increasing M3(L) to get higher BW. Mitigating Miller effect of M1 is the answer!

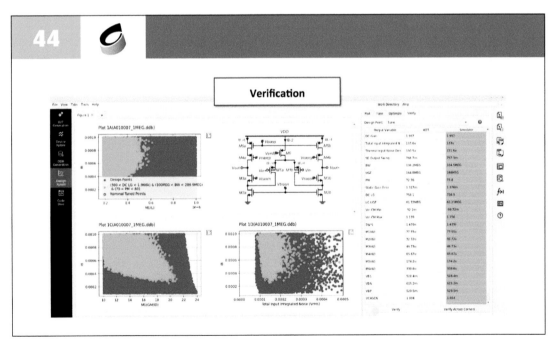

Do you want to make sure that ADT results are really simulator accurate? The verification process is just a single button click. The simulator results will be parsed and compared to ADT results.

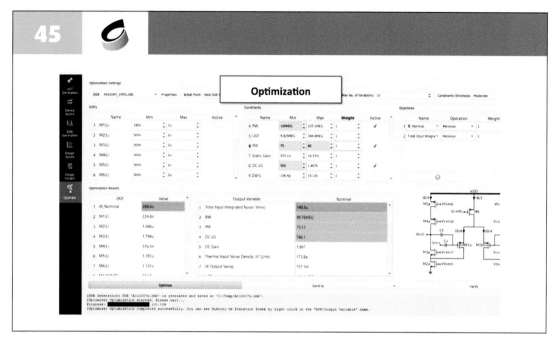

You can also optimize the same design using the optimization tab. Just put the constraints and objectives, and in a matter of seconds you will get the optimal design point!

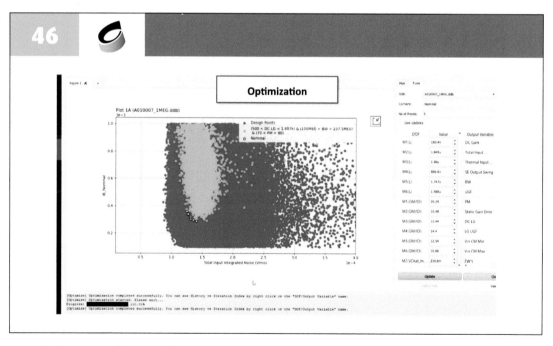

And what is even better is that you can visualize the optimization results in the design space. In this example we can see the pareto optimal front of power vs noise by varying the noise weight in the optimization process.

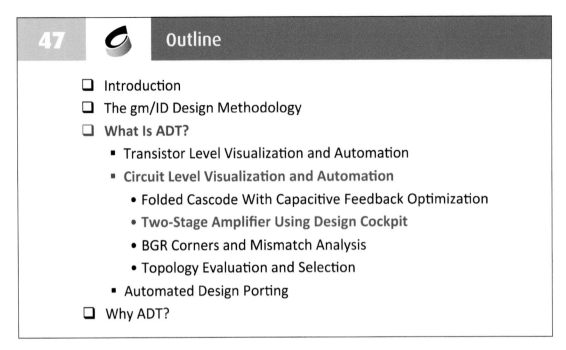

47   Outline

- ❏ Introduction
- ❏ The gm/ID Design Methodology
- ❏ **What Is ADT?**
  - ▪ Transistor Level Visualization and Automation
  - ▪ **Circuit Level Visualization and Automation**
    - • Folded Cascode With Capacitive Feedback Optimization
    - • **Two-Stage Amplifier Using Design Cockpit**
    - • BGR Corners and Mismatch Analysis
    - • Topology Evaluation and Selection
  - ▪ Automated Design Porting
- ❏ Why ADT?

Let's take another example to illustrate the powerful cockpit interface.

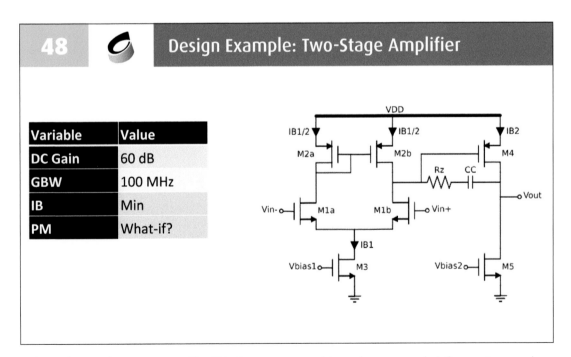

48   Design Example: Two-Stage Amplifier

Variable	Value
DC Gain	60 dB
GBW	100 MHz
IB	Min
PM	What-if?

This is the popular two-stage Miller OTA. Assume we want to explore many what-if scenarios, e.g., how much does the increase in phase margin affects the power consumption?

49

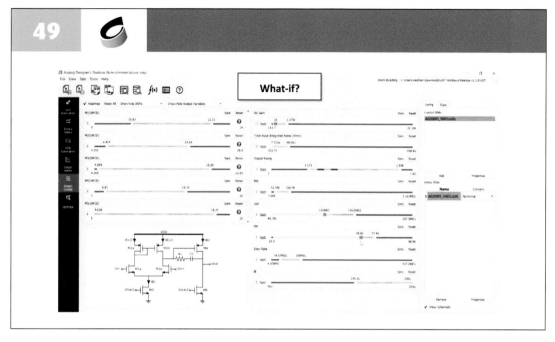

In the cockpit interface, you set constraints by simply dragging handles of a slider. You can see how this constraint is affecting all other specs and degrees of freedom. You can explore as many what-if scenarios as you want interactively.

50 **Outline**

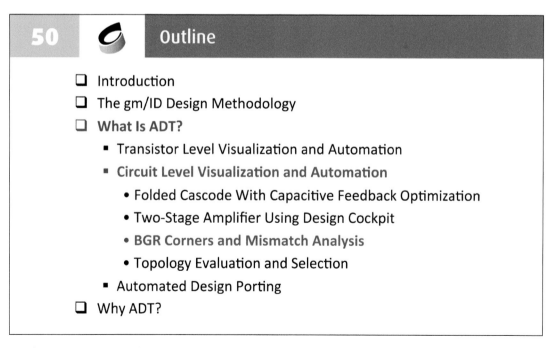

- ❑ Introduction
- ❑ The gm/ID Design Methodology
- ❑ What Is ADT?
  - ▪ Transistor Level Visualization and Automation
  - ▪ Circuit Level Visualization and Automation
    - • Folded Cascode With Capacitive Feedback Optimization
    - • Two-Stage Amplifier Using Design Cockpit
    - • BGR Corners and Mismatch Analysis
    - • Topology Evaluation and Selection
  - ▪ Automated Design Porting
- ❑ Why ADT?

Let's now turn to the ugly part of the design process, which is concerns and mismatch.

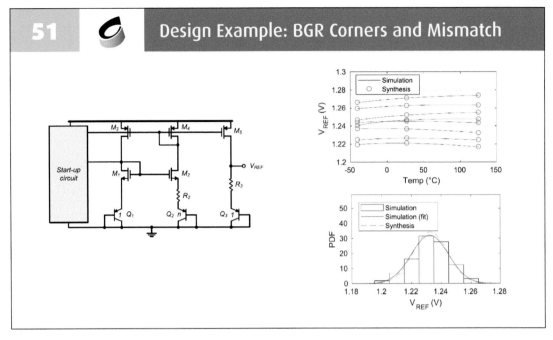

51 Design Example: BGR Corners and Mismatch

The good news is that we can build LUTs for corners and mismatch and get simulator-accurate results using ADT. This is an example a bandgap circuit that was published in "A. A. Youssef, B. Murmann and H. Omran, "Analog IC Design Using Precomputed Lookup Tables: Challenges and Solutions," in IEEE Access, vol. 8, pp. 134640-134652, 2020."

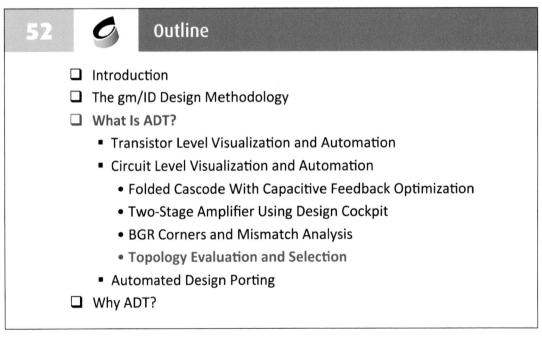

52 Outline

❑ Introduction
❑ The gm/ID Design Methodology
❑ What Is ADT?
  ▪ Transistor Level Visualization and Automation
  ▪ Circuit Level Visualization and Automation
    • Folded Cascode With Capacitive Feedback Optimization
    • Two-Stage Amplifier Using Design Cockpit
    • BGR Corners and Mismatch Analysis
    • Topology Evaluation and Selection
  ▪ Automated Design Porting
❑ Why ADT?

One more powerful thing you can do is to compare different topologies so that you can make informed design decisions when you select a topology for a set of specs.

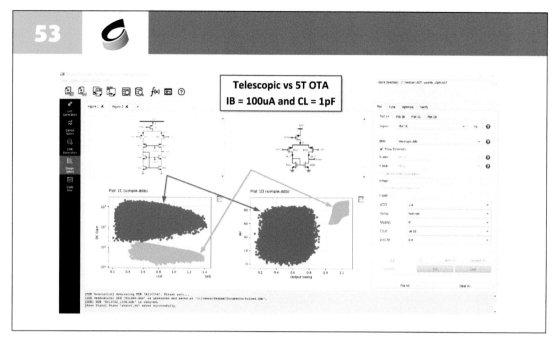

By painting the design space we can see that telescopic has higher gain, both have same UGF, DC gain of 5T OTA decreases at high UGF (small L), 5T OTA has higher swing and always good PM. Now you can easily make the right decision: should I just increase L or go to cascode?

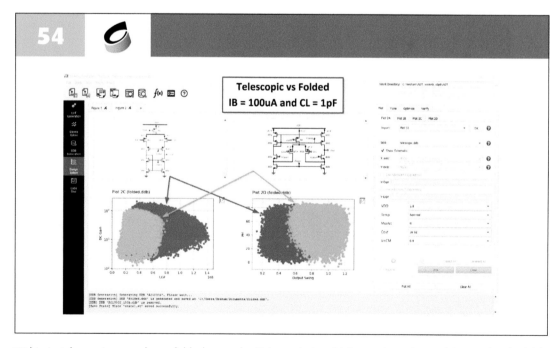

This is telescopic cascode vs folded cascode. Telescopic has higher gain and UGF (almost the double). Folded has higher output swing.

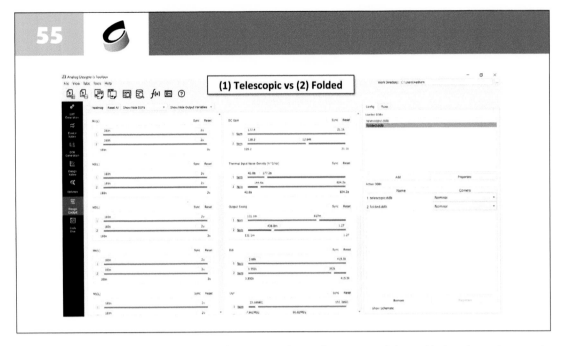

The topology comparison can be also done using the cockpit magic sliders. This is telescopic cascode vs folded cascode. Telescopic has higher gain and UGF (almost the double). Folded has higher output swing. Folded suffers from higher noise.

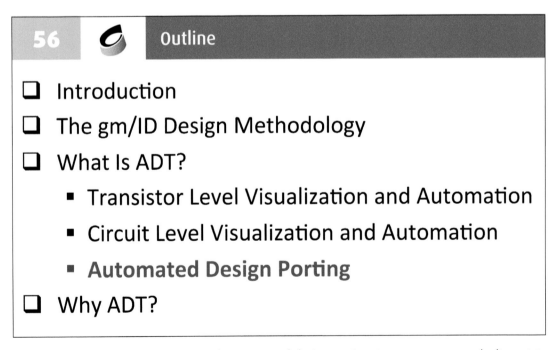

The third problem ADT is targeting is the automated design porting. In many cases, we don't want to create a new design from scratch, but we want to port an existing design from one technology to another.

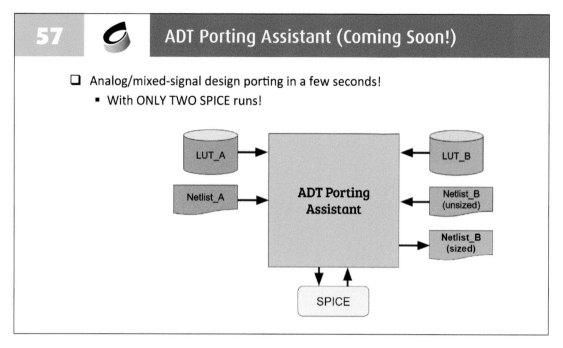

Using precomputed LUTs, ADT porting assistant can port a design from one technology to another with minimal number of SPICE runs. The idea is finding the "equivalent" transistor in the target technology and moving "transistor-wise".

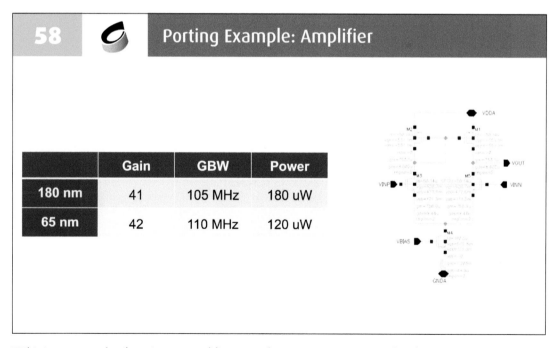

This is an example of porting an amplifier circuit from 180 nm to 65 nm. The algorithm can maintain the same gain and GBW. The power is reduced due to the reduced supply voltage.

59 **Porting Example: Dynamic Comparator**

	Delay	Power
130 nm	546 ps	8.9 uW
65 nm	342 ps	7.3 uW

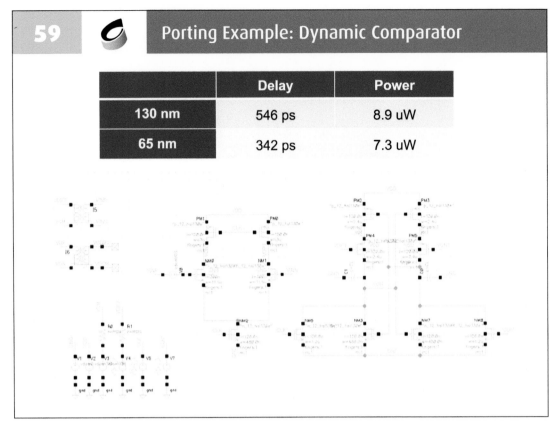

The porting algorithm can be also applied to digital and mixed-signal circuit. This is an example of dynamic comparator ported from 130 nm to 65 nm. The delay can power are reduced due to the smaller technology node.

60 **Outline**

❑ Introduction
❑ The gm/ID Design Methodology
❑ What Is ADT?
  ▪ Transistor Level Visualization and Automation
  ▪ Circuit Level Visualization and Automation
  ▪ Automated Design Porting
❑ **Why ADT?**

So why are we so excited about ADT? Why it is going to make a paradigm change?

61 Why ADT?

- ❑ Time savings
  - ▪ Analog building blocks designed in minutes instead of days
- ❑ Optimized power/performance/area
  - ▪ Global optimal design points
- ❑ Design insights
  - ▪ Paint the design space and visualize design trade-offs
- ❑ Process
  - ▪ Systematic, intuitive, and documentable design process

Because ADT can bring unique advantages to any professional designer and any student/researcher. Not only time savings and optimal quality, but also design insights and an intuitive design process.

62 What They Say About ADT

> ADT is indeed taking analog circuit design to a new level.
>
> – Pio Balmelli, Distinguished Engineer, Silicon Labs

> ADT provides analog designers with profound insight into the analog design process and guides them to understand, optimize and improve their designs.
>
> – Sherif Galal, Senior Director of Technology, Qualcomm

This is some of feedback we got about ADT from industry experts. ADT is currently used by several integrated circuit design companies from startups to big corporates.

63   What They Say About ADT

ADT is great.

– Wouter Serdijn, Professor and IEEE Fellow, Delft University

Very promising tool.

– Boris Murmann, Professor and IEEE Fellow, Stanford University

And this is some of the feedback from the academia. ADT is not just a great productivity tool for the industry, but also a great learning and teaching tool for the academia.

64   ADT Is AWARD WINNER!

DAC 2022 Innovator's Award!

ADT is also award winner! I received the prestigious DAC 2022 innovator's award in recognition of my work on ADT.

So we are trying to make a paradigm change in the analog design community. We want to change the 50-year old design methodology to a modern one that is fast, intuitive, optimized, and fun.

CHAPTER 06

Digital-Based Analog Processing for the IoT

Dr. Pedro Filipe Leite Correia De Toledo

Department of Electronics and Telecommunications (DET),
Politecnico di Torino, Italy
Graduate Program in Microelectronics (PGMICRO),
Universidade Federal of Rio Grande do Sul, Brazil
Synopsys, Portugal

1 **Summary**

- Introduction
  - IoT
  - Analog Design Challenges
  - Highly-Digital Design trend for analog/RF blocks
- Digital-Based Analog Differential Circuit
- Virtual Voltage Reference
- Bitstream D/A Conversion Techniques
  - Relaxation D/A Conversion
- Conclusion

In this talk I will present the results of several research activities, which have been carried out in the last years, and have been intended to translate analog functions into digital. I will show the results which we have obtained in translating into digital analog circuits like analog differential circuits and voltage references. Moreover, I will also discuss two bitstream digital-to-analog conversion techniques, since bitstream D/A conversion is very important in the implementation of analog functions in a digital way. Then I will draw some conclusions.

2 **Summary**

- **Introduction**
  - IoT
  - Analog Design Challenges
  - Highly-Digital Design trend for analog/RF blocks
- Digital-Based Analog Differential Circuit
- Virtual Voltage Reference
- Bitstream D/A Conversion Techniques
  - Relaxation D/A Conversion
- Conclusion

3 **Internet of Things**

Integrated circuits are embedded in objects to
- *Acquire*
- *Process*
- *Exchange*
useful information

Everyday life objects become **nodes** of a globally
interconnected **network**.

Let's start with the background of the Internet of Things: as we know, the Internet of Things is the vision of the world in which integrated circuits are embedded in everyday life objects so that to acquire, process and exchange useful information, thus acting in practice as nodes of a globally interconnected network.

4 **Internet of Things IC Requirements**

- **Tiny**
  - **From cm³ to *mm³*-scale, to be placed everywhere**

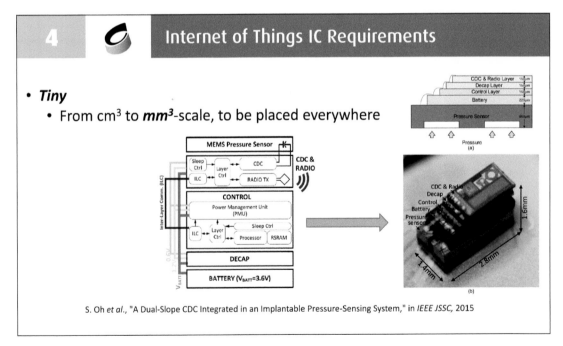

S. Oh *et al.*, "A Dual-Slope CDC Integrated in an Implantable Pressure-Sensing System," in *IEEE JSSC*, 2015

The implementation of the IoT paradigm raises very stringent constraints to integrated circuits, since IoT nodes need to be tiny, from the cubic centimeter down to the cubic millimeter scale, so that to be placed everywhere.

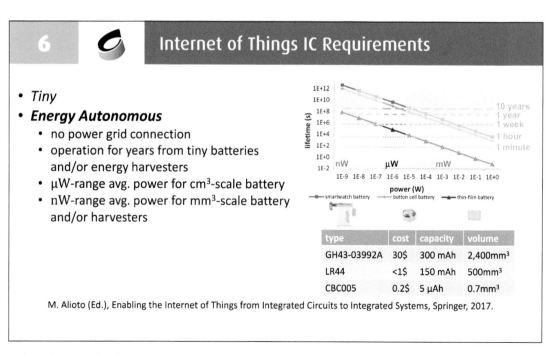

5 **Internet of Things IC Requirements**

- **Tiny**
  - From cm³ to $sub\text{-}mm^3$-scale, to be placed everywhere

Carrara, Sandro *"Body Dust: Miniaturized Highly-integrated Low Power Sensing for Remotely Powered Drinkable CMOS Bioelectronics."* 2018
S. Carrara, "Body Dust: Well Beyond Wearable and Implantable Sensors," in IEEE Sensors Journal, 2020
Y. Liu et al., "Bidirectional Bioelectronic Interfaces: System Design and Circuit Implications," in IEEE Solid-State Circuits Magazine, Spring 2020

The implementation of the IoT paradigm raises very stringent constraints to integrated circuits, since IoT nodes need to be tiny, from the cubic centimeter down to the cubic millimeter scale, so that to be placed everywhere.

6 **Internet of Things IC Requirements**

- **Tiny**
- **Energy Autonomous**
  - no power grid connection
  - operation for years from tiny batteries and/or energy harvesters
  - µW-range avg. power for cm³-scale battery
  - nW-range avg. power for mm³-scale battery and/or harvesters

type	cost	capacity	volume
GH43-03992A	30$	300 mAh	2,400mm³
LR44	<1$	150 mAh	500mm³
CBC005	0.2$	5 µAh	0.7mm³

M. Alioto (Ed.), Enabling the Internet of Things from Integrated Circuits to Integrated Systems, Springer, 2017.

Then, they need to be energy autonomous, since, of course, powering from the grid is not an option, and they need to be operated by tiny batteries or energy harvesters, which implies, in practice, that their average power consumption should be in the uW down to the nW range.

7 Internet of Things IC Requirements

- *Tiny*
- *Energy Autonomous*
- ***Always Connected***
- ***Reconfigurable***
  - Duty-cycled operation
  - Event-driven operation
  - Energy-quality scaling

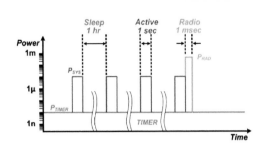

T. Jang *et al.*, "Circuit and System Designs of ULP Sensor Nodes With Illustration in a Miniaturized GNSS Logger for Position Tracking" in IEEE TCAS-I, 2017.

Considering that IoT nodes need to be always connected and have those stringent power constraints, an IoT node needs to be highly reconfigurable, so that to consume power only when strictly needed. For this reason, ICs need to be operated in duty-cycled or in event-driven mode and possibly, their performance needs to be dynamically traded off with the available power, thus performing energy-quality scaling.

8 Internet of Things IC Requirements

- *Tiny*
- *Energy Autonomous*
- *Always Connected*
- *Reconfigurable*
- ***Low Cost***
  - Pervasive technology: # of IoT nodes is expected to double in 4 years
  - The average cost of a node halves in 8 years
  - Technology costs to be minimized (mainstream CMOS)
  - Design costs to be minimized
  - Short time-to-market

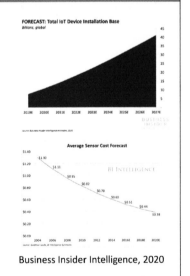

Business Insider Intelligence, 2020

Last, but not least, the cost of ICs for IoT applications need to be very very low, since this technology is expected to be pervasive - tens of billions of nodes are expected to be installed in the next few years - and, the cost of each single node needs to be minimized. It means that ICs should be implemented in mainstream CMOS technology and – most important – the design cost and effort need to be kept to a minimum.

9 **Summary**

- **Introduction**
  - **IoT**
  - **Analog Design Challenges**
  - **Highly-Digital Design trend for analog/RF blocks**
- **Digital-Based Analog Differential Circuit**
- **Virtual Voltage Reference**
- **Bitstream D/A Conversion Techniques**
  - **Relaxation D/A Conversion**
- **Conclusion**

In this talk I will present the results of several research activities, which have been carried out in the last years, and have been intended to translate analog functions into digital. I will show the results which we have obtained in translating into digital analog circuits like analog differential circuits and voltage references. Moreover, I will also discuss two bitstream digital to analog conversion techniques, since bitstream D/A conversion is very important in the implementation analog functions in a digital way. Then I will draw some conclusions.

10 **Analog Circuit Design Challenges**

- **Analog ICs do not take advantage of geometrical scaling**

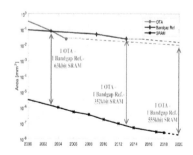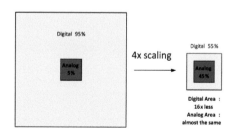

- **Almost no area reduction (limited by matching, noise…), relevant relative area contribution in scaled CMOS**

Even if analog ICs are and are going to be more and more the bottleneck in terms cost, performance and power of ICs in IoT applications, however, if we put in question whether we still need analog circuits in present day applications, the typical answer is that yes, we will always need analog circuits at least in interfaces towards an intrinsically analog physical world. Well, I will start my talk challenging this statement and, based on that, I will present several examples of a new design approach intended to implement analog functions by true, fully digital circuits.

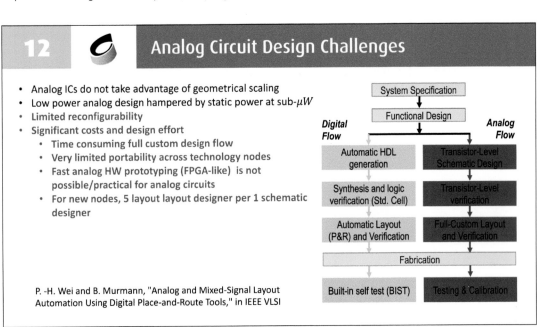

11 Analog Circuit Design Challenges

- Analog ICs do not take advantage of geometrical scaling

- Nanoscale devices show poor 'analog' features (low intrinsic gain g_m/g_o, leakage,...)
- Voltage scaling leads to SNR degradation
- Higher Layout challenges (Layout-Dependent Effects)

the time resolution is improving

P. R. Kinget, "Scaling analog circuits into deep nanoscale CMOS: Obstacles and ways to overcome them," IEEE CICC 2015

A. L. S. Loke et al., "Analog/mixed-signal design challenges in 7-nm CMOS and beyond," 2018 CICC

Even if analog ICs are and are going to be more and more the bottleneck in terms cost, performance and power of ICs in IoT applications, however, if we put in question whether we still need analog circuits in present day applications, the typical answer is that yes, we will always need analog circuits at least in interfaces towards an intrinsically analog physical world. Well, I will start my talk challenging this statement and, based on that, I will present several examples of a new design approach intended to implement analog functions by true, fully digital circuits.

12 Analog Circuit Design Challenges

- Analog ICs do not take advantage of geometrical scaling
- Low power analog design hampered by static power at sub-μW
- Limited reconfigurability
- Significant costs and design effort
  - Time consuming full custom design flow
  - Very limited portability across technology nodes
  - Fast analog HW prototyping (FPGA-like) is not possible/practical for analog circuits
  - For new nodes, 5 layout layout designer per 1 schematic designer

Digital Flow — *Analog Flow*

System Specification → Functional Design

Digital Flow	Analog Flow
Automatic HDL generation	Transistor-Level Schematic Design
Synthesis and logic verification (Std. Cell)	Transistor-Level verification
Automatic Layout (P&R) and Verification	Full-Custom Layout and Verification

Fabrication

| Built-in self test (BIST) | Testing & Calibration |

P. -H. Wei and B. Murmann, "Analog and Mixed-Signal Layout Automation Using Digital Place-and-Route Tools," in IEEE VLSI

Even if analog ICs are and are going to be more and more the bottleneck in terms cost, performance and power of ICs in IoT applications, however, if we put in question whether we still need analog circuits in present day applications, the typical answer is that yes, we will always need analog circuits at least in interfaces towards an intrinsically analog physical world. Well, I will start my talk challenging this statement and, based on that, I will present several examples of a new design approach intended to implement analog functions by true, fully digital circuits.

13   Summary

- **Introduction**
  - IoT
  - Analog Design Challenges
  - Highly-Digital Design trend for analog/RF blocks
- **Digital-Based Analog Differential Circuit**
- **Virtual Voltage Reference**
- **Bitstream D/A Conversion Techniques**
  - Relaxation D/A Conversion
- **Conclusion**

In this talk I will present the results of several research activities, which have been carried out in the last years, and have been intended to translate analog functions into digital. I will show the results which we have obtained in translating into digital analog circuits like analog differential circuits and voltage references. Moreover, I will also discuss two bitstream digital to analog conversion techniques, since bitstream D/A conversion is very important in the implementation analog functions in a digital way. Then I will draw some conclusions.

14   Highly-Digital Design trend for analog/RF blocks

In view of the limitations of analog circuits in nanoscale CMOS, **all digital** and/or digital **intensive** replacements of IC cells commonly implemented by analog techniques have been extensively investigated

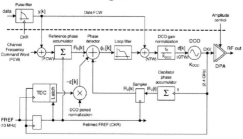

All-digital PLL

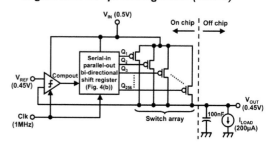

Digital Low Dropout Regulator (DLDO)

R. B. Staszewski *et al.*, "All-digital TX frequency synthesizer and discrete-time receiver for Bluetooth radio in 130-nm CMOS," in *IEEE JSSC*, Dec. 2004

Y. Okuma *et al.*, "0.5-V input digital LDO with 98.7% current efficiency and 2.7µA quiescent current in 65 nm CMOS", *IEEE CICC*, 2010.

In view of the limitations of analog integrated circuits, in recent years there has been a strong trend toward the implementation of traditionally analog blocks by all-digital or digital intensive replacements. For instance, in 2004, the all-digital PLL has been proposed and in 2010 the digital LDO has been introduced and both these cells have become quite popular in recent years.

15 Highly-Digital Design trend for analog/RF blocks

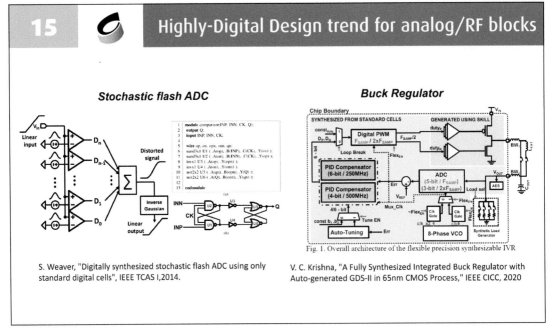

Stochastic flash ADC

Buck Regulator

Fig. 1. Overall architecture of the flexible precision synthesizable IVR

S. Weaver, "Digitally synthesized stochastic flash ADC using only standard digital cells", IEEE TCAS I,2014.

V. C. Krishna, "A Fully Synthesized Integrated Buck Regulator with Auto-generated GDS-II in 65nm CMOS Process," IEEE CICC, 2020

In view of the limitations of analog integrated circuits, in recent years there has been a strong trend toward the implementation of traditionally analog blocks by all-digital or digital intensive replacements. For instance, in 2004, the all-digital PLL has been proposed and in 2010 the digital LDO has been introduced and both these cells have become quite popular in recent years.

16 Highly-Digital Design trend for analog/RF blocks

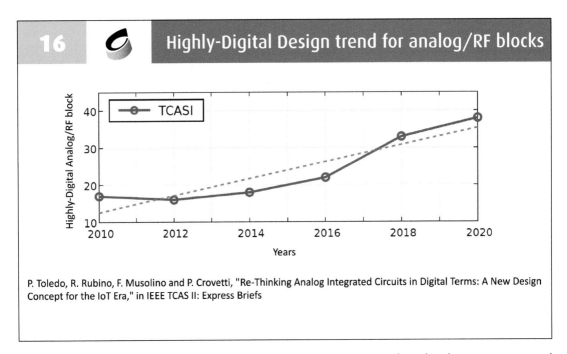

P. Toledo, R. Rubino, F. Musolino and P. Crovetti, "Re-Thinking Analog Integrated Circuits in Digital Terms: A New Design Concept for the IoT Era," in IEEE TCAS II: Express Briefs

In view of the limitations of analog integrated circuits, in recent years there has been a strong trend toward the implementation of traditionally analog blocks by all-digital or digital intensive replacements. For instance, in 2004, the all-digital PLL has been proposed and in 2010 the digital LDO has been introduced and both these cells have become quite popular in recent years.

17 Analog Processing by Digital Means

- **Target**: processing arbitrary band-limited input signal(s) (*v* and/or *i*) to get band-limited output signal(s) (*v* and/or *i*) at pre-determined resolution by a true **digital circuit.**
- **Digital circuit**
  - Information internally encoded as binary signals (0/1), synchronous or asynchronous operation (can include e.g., combinational and sequential logic, registers, memories, FSMs...).
  - <u>NO digital gates used as analog circuits</u> (e.g. CMOS inverter → inverting amplifier)
- Can include *minimal, non-critical,* **passive networks** (e.g. *RC* filters, volt. divider) *if necessary*.

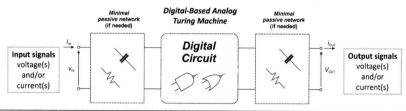

How to bring the "digital revolution" to analog interfaces? Shannon's results suggest that it is feasible, but do not provide a practical indication on how to do that, as it often happens with theoretical results. In this talk I am going to suggest some approaches which are intended to translate typical analog circuits into digital. What I mean with "translate analog circuits into digital"? Well, I mean to develop true digital circuits, which are suitable to process arbitrary, band-limited input signals (voltages and/or currents) to get any arbitrary band-limited output signal (voltage, current) at a fixed degree of accuracy. And with *digital circuit*, I mean a circuit in which information is processed and encoded internally in digital signals, which can take just to values ("1" or "0"), circuits which can be synchronous or asynchronous, may include combinational or sequential logic, registers, memories, FSMs..., and in which digital gates are <u>NOT</u> used as analog circuits (I am not talking about circuits in which digital gates, e.g. CMOS inverters, are used as analog amplifiers). We know that a CMOS inverter can be used as an inverting amplifier, but this is not what we are talking about. We can just accept that in front of the inputs and after the output of our digital circuits we may have minimal, non-critical passives, whenever it is strictly necessary. By so doing, we can develop a kind of paradigm for digital-based analog processing, a kind of digital-based Turing machine, which is intended to process within a target degree of accuracy and bandwidth limitation *any* kind of analog signal, so that to provide any kind of analog output.

18 Summary

In this talk I will present the results of several research activities, which have been carried out in the last years, and have been intended to translate analog functions into digital. I will show the results which we have obtained in translating into digital analog circuits like analog differential circuits and voltage references. Moreover, I will also discuss two bitstream digital to analog conversion techniques, since bitstream D/A conversion is very important in the implementation analog functions in a digital way. Then I will draw some conclusions.

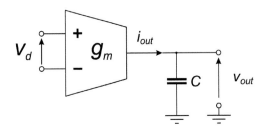

19 — Operational Transconductance Amplifier (OTA)

Key building block in analog electronics
Mostly used as error amplifier in **negative feedback** configurations

Analog Description

$$i_{out} = g_m v_d$$

$$v_{out} = \frac{g_m}{C} \int_0^t v_d(t')dt'$$

i_{out}, v_{out} insensitive to $v_{CM} = \frac{v^+ + v^-}{2}$

Can we translate it into digital?

Let's consider an operational transconductance amplifier (OTA), which is a key building block in analog electronics, mostly used as an error amplifier in negative feedback configurations. As we know, this circuit is expected to provide an output current which is proportional to the input voltage across its input terminals, i.e. $v^+ - v^-$, and is insensitive to the common-mode input voltage, i.e. $\frac{v^+ + v^-}{2}$. This circuit is very often considered with a capacitive load, so that the output voltage of this circuit is basically proportional to the integral of the input differential voltage. Now the question is: can we translate an OTA into digital?

20 — Digital-Based Analog Differential Circuit

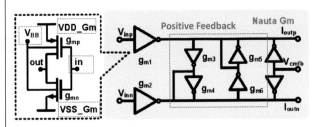

Nauta OTA

☺ Inverter based
☺ High frequency
☹ Digital inverter used as an analog amplifier (amplitude-domain)
☹ Static power

$$A_{dc} = \frac{g_{m1}}{1/r_{out} + g_{m4} - g_{m3}}$$

B. Nauta, "A CMOS transconductance-C filter technique for very high frequencies," in *IEEE JSSC*, Feb. 1992

L. Lv, A. Jain "A 0.4-V Gm–C Proportional-Integrator-Based Continuous-Time DSM With 50-kHz BW and 74.4-dB SNDR," in IEEE JSSC 2018.

Yet, this is not the first attempt to develop a digital-based, a digital friendly implementation of an OTA, we can recall, the Nauta circuit, which is an inverter-based topology and is also suitable to operate at high frequency, but this is definitely not a digital circuit since inverters are used here as analog amplifiers and the information is processed in the analog domain (moreover, the circuit is not suitable to IoT applications since it draws a significant power).

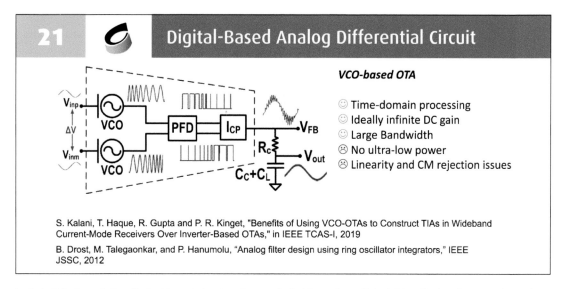

21 **Digital-Based Analog Differential Circuit**

VCO-based OTA

☺ Time-domain processing
☺ Ideally infinite DC gain
☺ Large Bandwidth
☹ No ultra-low power
☹ Linearity and CM rejection issues

S. Kalani, T. Haque, R. Gupta and P. R. Kinget, "Benefits of Using VCO-OTAs to Construct TIAs in Wideband Current-Mode Receivers Over Inverter-Based OTAs," in IEEE TCAS-I, 2019

B. Drost, M. Talegaonkar, and P. Hanumolu, "Analog filter design using ring oscillator integrators," IEEE JSSC, 2012

Yet, this is not the first attempt to develop a digital-based, a digital friendly implementation of an OTA, we can recall, the Nauta circuit, which is an inverter-based topology and is also suitable to operate at high frequency, but this is definitely not a digital circuit since inverters are used here as analog amplifiers and the information is processed in the analog domain (moreover, the circuit is not suitable to IoT applications since it draws a significant power).

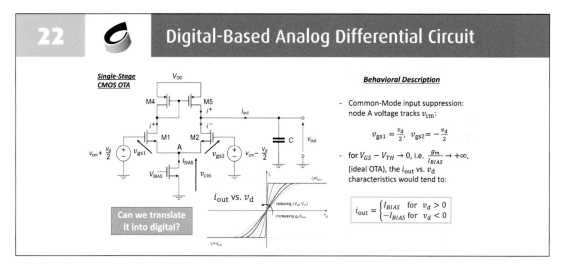

22 **Digital-Based Analog Differential Circuit**

Single-Stage CMOS OTA

Behavioral Description

- Common-Mode input suppression: node A voltage tracks v_{cm}:

$$v_{gs1} = \frac{v_d}{2}, \; v_{gs2} = -\frac{v_d}{2}$$

- for $V_{GS} - V_{TH} \to 0$, i.e. $\frac{gm}{I_{BIAS}} \to +\infty$, (ideal OTA), the i_{out} vs. v_d characteristics would tend to:

$$i_{out} = \begin{cases} I_{BIAS} & \text{for } v_d > 0 \\ -I_{BIAS} & \text{for } v_d < 0 \end{cases}$$

i_{out} vs. v_d

Can we translate it into digital?

Now, let's consider a different approach. Let's start from a simple traditional implementation of a CMOS OTA and let's discuss if we can translate it into a digital circuit. For this purpose, rather than starting from small-signal analysis, which is typical of analog circuits, let's start from a behavioral description of this circuit, as a starting point to get a digital implementation. Well, we can observe that the input differential pair in this circuit provides common-mode input signal suppression: we have the CM signal component is tracked by the voltage of the common-source node A, and such voltage is actually subtracted from the external inputs in the Gate Source voltages of the input devices (v_gs+=v+-vcm=vd/2 and v_gs-= v- -vcm=-vd/2), so the control voltages of the input devices are CM-voltage independent and they translate these voltages into a drain current which is just related to the differential mode input. The current mirror then sums the two differential current contributions with the correct sign to get the output current. If we consider the id(vd) characteristics of this stage, for gm/Ibias tending to infinity (which cannot be achieved in practice in CMOS, but would be highly desirable in an OTA to have a very large transconductance gain at low current), we have that the behavior of this circuit can be described in this simple way: the output current is +Ibias, if the input differential voltage is positive, whereas it is -Ibias if the input differential voltage is negative.

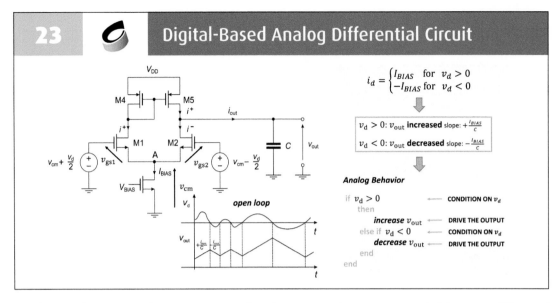

 ooking at the output voltage, we can say that the output voltage is increased if the input differential voltage is positive, whereas it is decreased when the input differential voltage is negative. The positive and the negative slopes of the output voltage are the same in this very simple description. By the way, even if they are not the same, even if they are not constant, this is not a big concern, as far as these slopes are higher than the slew rate of the output signal, when the OTA is employed in a negative feedback configuration. Based on that, we can say that the behavior of the OTA can be described by this simple analog pseudo-code: if the differential voltage is greater than zero, then increase vout, else, if the differential voltage is less than zero, decrease vout. As a consequence, to implement an OTA, we have to do two things: first, we have to check a condition on the input differential voltage, and then we have to drive the output accordingly.

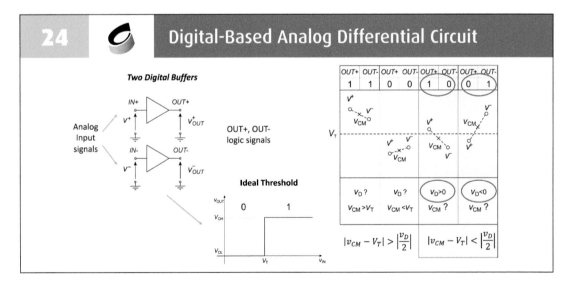

ow, let's suppose to have two digital buffers, with an ideal threshold behavior (the output is at zero if the input is below the trip point voltage, and it is at one if the input voltage is above the trip point VT) and let's suppose to apply the input signals of our OTA to these two digital buffers. Let's consider the digital output of these gates and let's discuss if we can implement the function of an OTA based on the digital values of these signals. Of course, these two signals can take 4 different values (11, 00, 10, 01). We can immediately observe that if OUT+=1 and OUT-=0, we can conclude that vd>0. In the same way, if we have that OUT+=0 and OUT-=1, we can immediately conclude that vd<0.

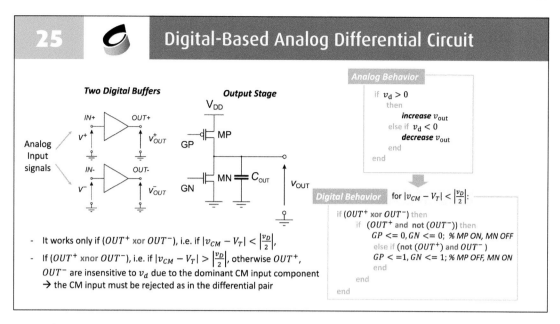

25 **Digital-Based Analog Differential Circuit**

Two Digital Buffers

Output Stage

Analog Behavior

if $v_d > 0$
then
increase v_{out}
else if $v_d < 0$
decrease v_{out}
end
end

Digital Behavior for $|v_{CM} - V_T| < |\frac{v_D}{2}|$:

if $(OUT^+$ xor $OUT^-)$ then
if $(OUT^+$ and not $(OUT^-))$ then
GP <= 0, GN <= 0; % MP ON, MN OFF
else if (not (OUT^+) and OUT^-)
GP < =1, GN <= 1; % MP OFF, MN ON
end
end
end

- It works only if $(OUT^+$ xor $OUT^-)$, i.e. if $|v_{CM} - V_T| < |\frac{v_D}{2}|$,
- If $(OUT^+$ xnor $OUT^-)$, i.e. if $|v_{CM} - V_T| > |\frac{v_D}{2}|$, otherwise OUT^+, OUT^- are insensitive to v_d due to the dominant CM input component
→ the CM input must be rejected as in the differential pair

Based on this information, we can drive an output stage, which can be a three-state buffer, so that to increase or decrease the output voltage. In other words, if OUT+=1 and OUT-=0, we can turn on the pull up device so that to source a positive output current, whereas if OUT+=0 and OUT-=1, we can turn on the pull down device so that to draw current from the output and decrease the output voltage. Unfortunately, this works only if OUT+, OUT- are (0,1) or (1,0),

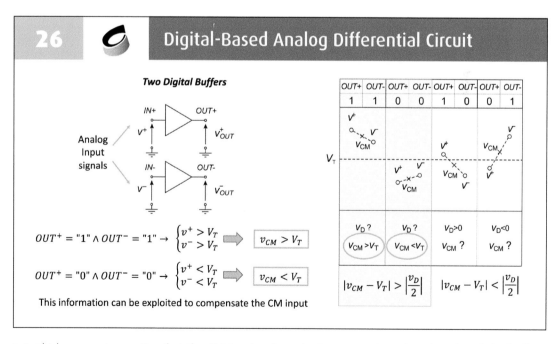

26 **Digital-Based Analog Differential Circuit**

Two Digital Buffers

Analog Input signals

$OUT^+ = "1" \wedge OUT^- = "1" \rightarrow \begin{cases} v^+ > V_T \\ v^- > V_T \end{cases}$ ⟹ $v_{CM} > V_T$

$OUT^+ = "0" \wedge OUT^- = "0" \rightarrow \begin{cases} v^+ < V_T \\ v^- < V_T \end{cases}$ ⟹ $v_{CM} < V_T$

This information can be exploited to compensate the CM input

OUT+	OUT-	OUT+	OUT-	OUT+	OUT-	OUT+	OUT-
1	1	0	0	1	0	0	1
V_D?		V_D?		$V_D>0$		$V_D<0$	
$V_{CM}>V_T$		$V_{CM}<V_T$		V_{CM} ?		V_{CM} ?	
$\|v_{CM} - V_T\| > \|\frac{v_D}{2}\|$				$\|v_{CM} - V_T\| < \|\frac{v_D}{2}\|$			

Which means in practice that the CM input voltage is close enough to the trip point of the buffers, closer than vd/2, whenever vcm is not so close to the trip point of the buffers, we have OUT+,OUT- (1,1) or (0,0). In this case, the digital outputs of the buffers do not provide any useful information on the input differential voltage. By the way, these outputs provide useful information on the common-mode input voltage: if OUT+,OUT- = (1,1) it means that Vcm > VT, whereas if OUT+,OUT- = (0,0), it means that the CM input voltage is below the trip point. Based on this information, we can add a compensation signal to the input of the buffers so that to make it closer to VT, to enforce the operation of the circuit as an OTA.

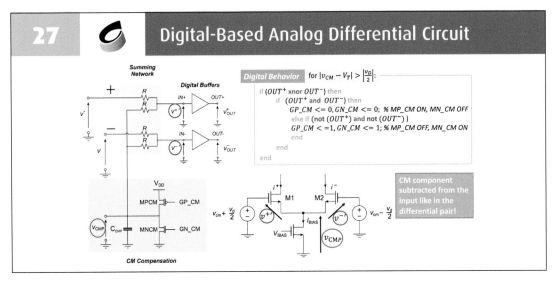

In practice, this behavior can be obtained by adding to v+, v- a compensation signal VCMP, generated by a 3-state buffer according to the digital values of OUT+ and OUT-. In other words, whenever OUT+ and OUT- are both 0, which means that the common-mode input voltage is too low, below the trip points of the buffers, the pull-up transistor MPCM is turned on to increase the common-mode compensation voltage. By contrast, if both OUT+ and OUT- are high, the pull-down transistor MNCM is turned on to decrease common-mode compensation voltage added to the input. The behavior of this network is analogous to the behavior of a differential pair, in which the source node voltage, which tracks the CM input variations, is subtracted from the external inputs. So, we can say this is the digital implementation of a differential pair.

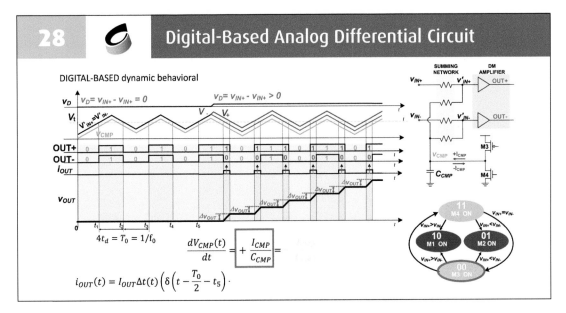

Here we can see the internal nodes of Digital-Based OTA when it is working with and without input differential signal. When there is no diff. signal, the digital-based OTA oscillates like it were a relaxation oscillator: charging and discharging the capacitor CCMP. In this way, nothing happens at the output as long as only the common mode signal is equal. It means that the common mode signal is being rejected. At the moment that there is any kind of differential signal, there is going to be a different initial condition, which is proportional to the input signal, in the charging and discharging process. In this way, there is a phase shift passing through the buffer generating a pulse width proportional to the differential input signal. Then, these pulses can be caught by the output signal amplifying the differential signal

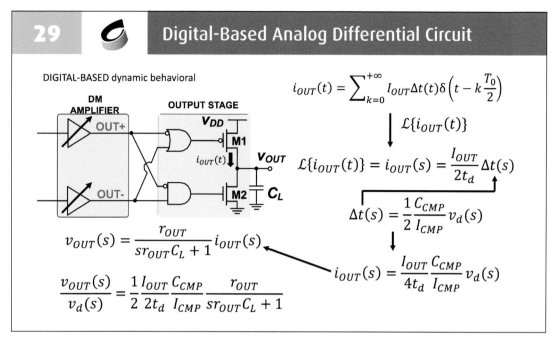

29 — **Digital-Based Analog Differential Circuit**

DIGITAL-BASED dynamic behavioral

$$i_{OUT}(t) = \sum_{k=0}^{+\infty} I_{OUT}\Delta t(t)\delta\left(t - k\frac{T_0}{2}\right)$$

$$\mathcal{L}\{i_{OUT}(t)\}$$

$$\mathcal{L}\{i_{OUT}(t)\} = i_{OUT}(s) = \frac{I_{OUT}}{2t_d}\Delta t(s)$$

$$\Delta t(s) = \frac{1}{2}\frac{C_{CMP}}{I_{CMP}}v_d(s)$$

$$i_{OUT}(s) = \frac{I_{OUT}}{4t_d}\frac{C_{CMP}}{I_{CMP}}v_d(s)$$

$$v_{OUT}(s) = \frac{r_{OUT}}{sr_{OUT}C_L + 1}i_{OUT}(s)$$

$$\frac{v_{OUT}(s)}{v_d(s)} = \frac{1}{2}\frac{I_{OUT}}{2t_d}\frac{C_{CMP}}{I_{CMP}}\frac{r_{OUT}}{sr_{OUT}C_L + 1}$$

In this slide, we can see how the transfer function can be calculated to the Digital-Based OTA.

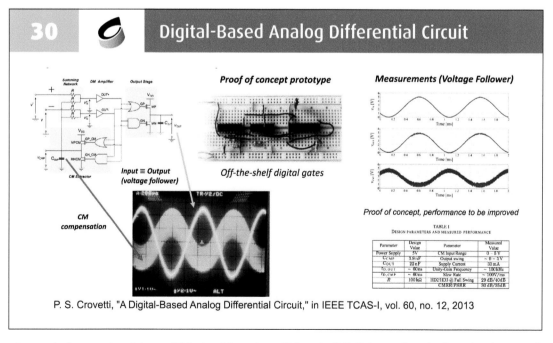

30 — **Digital-Based Analog Differential Circuit**

Proof of concept prototype

Measurements (Voltage Follower)

Input ≡ Output (voltage follower)

Off-the-shelf digital gates

CM compensation

Proof of concept, performance to be improved

TABLE I
DESIGN PARAMETERS AND MEASURED PERFORMANCE

Parameter	Design Value	Parameter	Measured Value
Power Supply	5V	CM Input Range	0 – 5V
C_CMP	3.9nF	Output swing	~ 0 – 5V
C_OUT	22 nF	Supply Current	33 mA
t_D,OUT	~ 60ns	Unity-Gain Frequency	~ 100kHz
t_D,CMP	~ 60ns	Slew Rate	~ 100V/ms
R	100 kΩ	HD2/HD3 @ Full Swing	29 dB/40dB
		CMRR/PSRR	30 dB/35dB

P. S. Crovetti, "A Digital-Based Analog Differential Circuit," in IEEE TCAS-I, vol. 60, no. 12, 2013

A proof-of-concept prototype of this circuit based on off-the-shelf digital gates has also been implemented. The circuit has been tested in the voltage follower configuration and it works properly: the output signal follows the input signal, and the CM compensation signal is generated so that to keep the CM input voltage close to VDD/2. The performance is not outstanding, but it proves the concept.

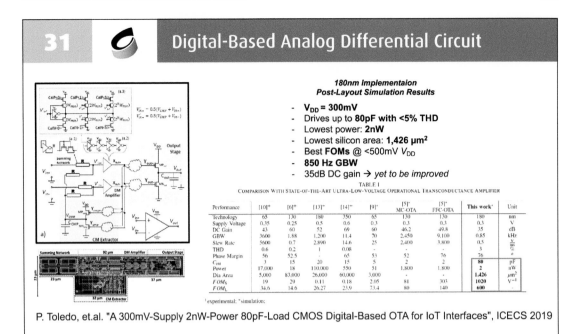

31 **Digital-Based Analog Differential Circuit**

180nm Implementaion
Post-Layout Simulation Results

- **V_{DD} = 300mV**
- Drives up to **80pF with <5% THD**
- Lowest power: **2nW**
- Lowest silicon area: **1,426 μm²**
- Best **FOMs** @ <500mV V_{DD}
- **850 Hz GBW**
- 35dB DC gain → *yet to be improved*

TABLE I
COMPARISON WITH STATE-OF-THE-ART ULTRA-LOW-VOLTAGE OPERATIONAL TRANSCONDUCTANCE AMPLIFIER

Performance	[10]⁺	[6]⁺	[13]⁻	[14]⁻	[9]*	[5]* MC-OTA	[5]* FFC-OTA	This work*	Unit
Technology	65	130	180	350	65	130	130	180	nm
Supply Voltage	0.35	0.25	0.5	0.6	0.3	0.3	0.3	0.3	V
DC Gain	43	60	52	69	60	46.2	49.8	35	dB
GBW	3600	1.88	1,200	11.4	70	2,450	9,100	0.85	kHz
Slew Rate	5600	0.7	2,890	14.6	25	2,400	3,800	0.5	V/ms
THD	0.6	0.2	1	0.08	-	-	-	3	%
Phase Margin	56	52.5	-	65	53	52	76	76	°
C_{out}	3	15	20	15	5	2	2	80	pF
Power	17,000	18	110,000	550	51	1,800	1,800	2	nW
Dia Area	5,000	83,000	26,000	60,000	3,000	-	-	1,426	μm²
FOM_S	19	29	0.11	0.18	2.05	81	303	1020	V⁻¹
FOM_L	34.6	14.6	26.27	23.9	73.4	80	140	600	-

⁺ experimental; * simulation;

P. Toledo, et.al. "A 300mV-Supply 2nW-Power 80pF-Load CMOS Digital-Based OTA for IoT Interfaces", ICECS 2019

More recently, the digital OTA has been implemented in 180nm by digital standard cells and the simulations reveal the potential of this technique in IoT applications: the circuit works at Vdd down to 300mVs, it is suitable to drive large capacitive loads up to 80pF with less than 5% THD and, most important, it draws just 2nW power. Its silicon area is just 1,462um^2 in 180nm CMOS and it achieves among the best FOMs compared to analog solutions working at Vdd below 500mV. What is even more interesting, we have also simulated this circuit in 40nm and porting the circuit from 180nm to 40nm both the FOM_S and FOM_L of this digital OTA are significantly improved, showing that unlike analog circuits, this digital OTA takes advantage of scaling. Also in this sense it can be regarded as a true digital circuit.

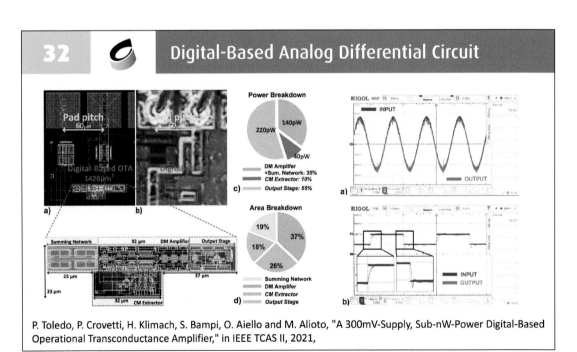

32 **Digital-Based Analog Differential Circuit**

P. Toledo, P. Crovetti, H. Klimach, S. Bampi, O. Aiello and M. Alioto, "A 300mV-Supply, Sub-nW-Power Digital-Based Operational Transconductance Amplifier," in IEEE TCAS II, 2021,

The circuit has also been fabricated on silicon and, based on preliminary measurement results, the circuit works as expected and the measured performance fully confirms what was observed in simulations.

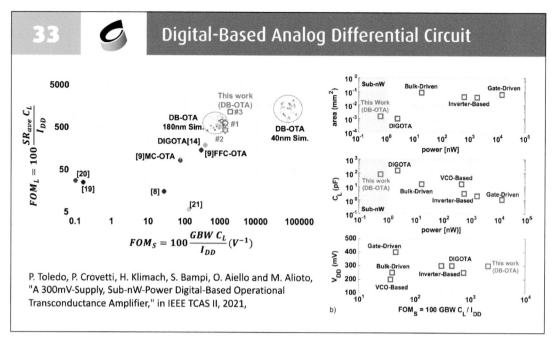

33 **Digital-Based Analog Differential Circuit**

P. Toledo, P. Crovetti, H. Klimach, S. Bampi, O. Aiello and M. Alioto, "A 300mV-Supply, Sub-nW-Power Digital-Based Operational Transconductance Amplifier," in IEEE TCAS II, 2021,

The circuit has also been fabricated on silicon and, based on preliminary measurement results, the circuit works as expected and the measured performance fully confirms what was observed in simulations.

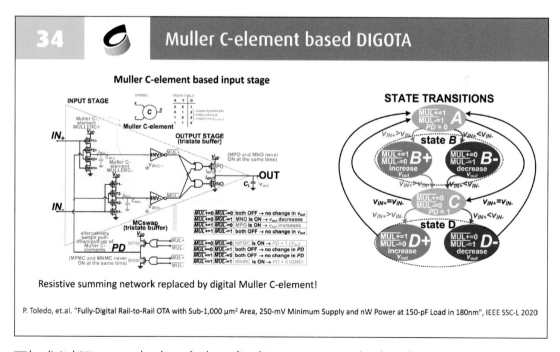

34 **Muller C-element based DIGOTA**

Resistive summing network replaced by digital Muller C-element!

P. Toledo, et.al. "Fully-Digital Rail-to-Rail OTA with Sub-1,000 µm² Area, 250-mV Minimum Supply and nW Power at 150-pF Load in 180nm", IEEE SSC-L 2020

The digital OTA concept has been further refined in more recent work, where the input summing network, which is not very convenient for integration on silicon, has been replaced by digital elements, i.e. Muller C-elements, which prove to be suitable to implement the same function. The behavior of this circuit can be fully described in terms of a state transition graph, as a finite state machine.

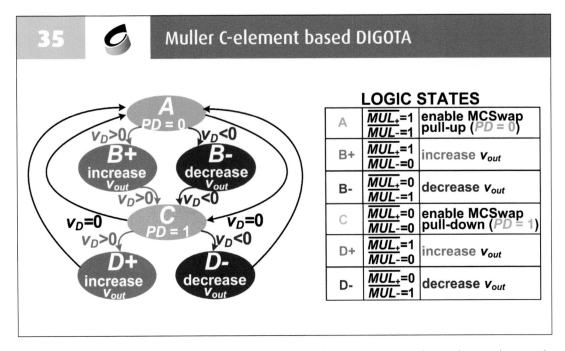

35 — **Muller C-element based DIGOTA**

	LOGIC STATES	
A	$\overline{MUL_+}=1$ $\overline{MUL_-}=1$	enable MCSwap pull-up (PD = 0)
B+	$\overline{MUL_+}=1$ $MUL_-=0$	increase v_{out}
B-	$\overline{MUL_+}=0$ $\overline{MUL_-}=1$	decrease v_{out}
C	$MUL_+=0$ $\overline{MUL_-}=0$	enable MCSwap pull-down (PD = 1)
D+	$MUL_+=1$ $MUL_-=0$	increase v_{out}
D-	$MUL_+=0$ $MUL_-=1$	decrease v_{out}

This shows the truth table of DIGOTA. In the table on the right, it summarizes what each state does inside of DIGOTA operation.

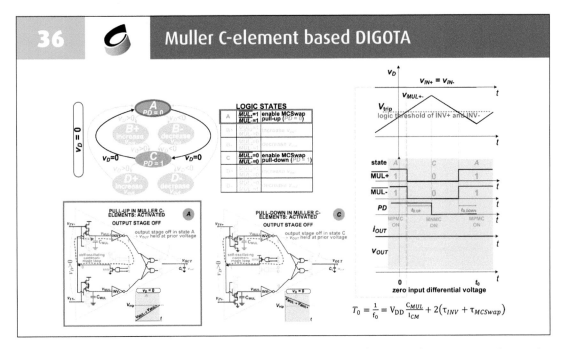

36 — **Muller C-element based DIGOTA**

$$T_0 = \frac{1}{f_0} = V_{DD} \frac{C_{MUL}}{I_{CM}} + 2(\tau_{INV} + \tau_{MCSwap})$$

When there is no input differential signal, the DIGOTA oscillates between the states A and C. At the same way, These states correspond to the 00 and 11, which do not trigger the output stage.

37 Muller C-element based DIGOTA

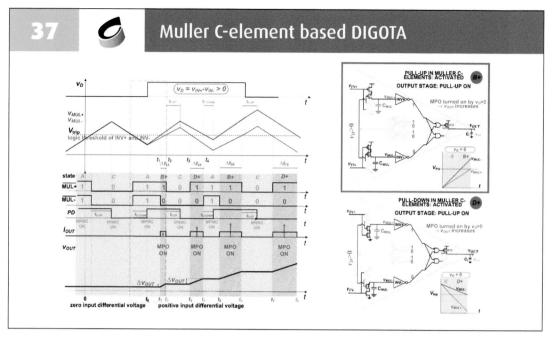

At the moment we get a diff. signal, the parasitic capacitance at the output stage of muller-C are both charging or discharging with different rates generating the same effect as before. In this way, the train of pulsed proportional to the input can be caught by the output stage.

38 Muller C-element based DIGOTA

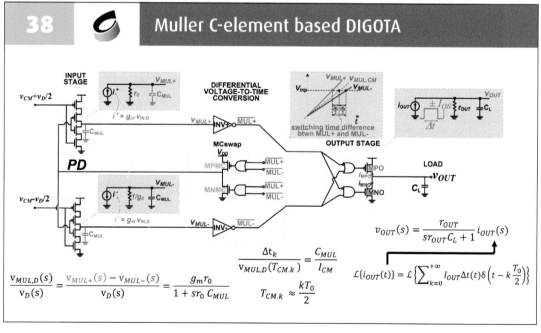

Using a similar approach as before, we can calculate the transfer function of DIGOTA.

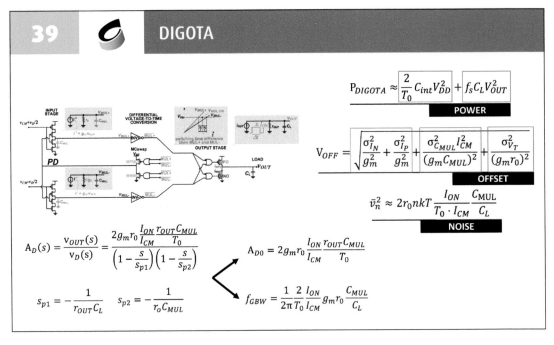

39　　　　　**DIGOTA**

$$P_{DIGOTA} \approx \frac{2}{T_0} C_{int} V_{DD}^2 + f_s C_L V_{OUT}^2$$

POWER

$$V_{OFF} = \sqrt{\frac{\sigma_{I_N}^2}{g_m^2} + \frac{\sigma_{I_P}^2}{g_m^2} + \frac{\sigma_{C_{MUL}}^2 I_{CM}^2}{(g_m C_{MUL})^2} + \frac{\sigma_{V_T}^2}{(g_m r_0)^2}}$$

OFFSET

$$\bar{v}_n^2 \approx 2r_0 nkT \frac{I_{ON}}{T_0 \cdot I_{CM}} \frac{C_{MUL}}{C_L}$$

NOISE

$$A_D(s) = \frac{v_{OUT}(s)}{v_D(s)} = \frac{2g_m r_0 \frac{I_{ON}}{I_{CM}} \frac{r_{OUT} C_{MUL}}{T_0}}{\left(1 - \frac{s}{s_{p1}}\right)\left(1 - \frac{s}{s_{p2}}\right)}$$

$$A_{D0} = 2g_m r_0 \frac{I_{ON}}{I_{CM}} \frac{r_{OUT} C_{MUL}}{T_0}$$

$$s_{p1} = -\frac{1}{r_{OUT} C_L} \qquad s_{p2} = -\frac{1}{r_o C_{MUL}}$$

$$f_{GBW} = \frac{1}{2\pi} \frac{2}{T_0} \frac{I_{ON}}{I_{CM}} g_m r_0 \frac{C_{MUL}}{C_L}$$

B ased on that, we got a second-order system.

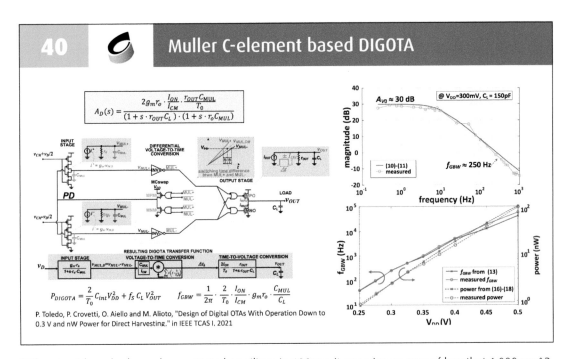

40　　　　　**Muller C-element based DIGOTA**

$$A_D(s) = \frac{2g_m r_o \cdot \frac{I_{ON}}{I_{CM}} \cdot \frac{r_{OUT} C_{MUL}}{T_0}}{(1 + s \cdot r_{OUT} C_L) \cdot (1 + s \cdot r_o C_{MUL})}$$

$A_{V0} \approx 30 \text{ dB}$　@ V_{DD}=300mV, C_L = 150pF

$f_{GBW} \approx 250 \text{ Hz}$

— (10)-(11)　·○· measured

$$P_{DIGOTA} = \frac{2}{T_0} C_{int} V_{DD}^2 + f_s C_L V_{OUT}^2 \qquad f_{GBW} = \frac{1}{2\pi} \cdot \frac{2}{T_0} \frac{I_{ON}}{I_{CM}} \cdot g_m r_o \cdot \frac{C_{MUL}}{C_L}$$

f_{GBW} from (13)　·○· measured f_{GBW}　·×· power from (16)-(18)　·○· measured power

P. Toledo, P. Crovetti, O. Aiello and M. Alioto, "Design of Digital OTAs With Operation Down to 0.3 V and nW Power for Direct Harvesting," in IEEE TCAS I, 2021

The circuit has also been demonstrated on silicon in 180nm, it occupies an area of less that 1,000 um^2, operates to VDD<300mV driving up to 150pF with a power in the nW range which has been proven to be suitable for direct operation from an energy harvester (a photodiode with less than 7mm^2 area) and achieve state of the art FOMs among OTAs working below 500mV. This confirms that this circuit is suitable to address the challenges of IoT applications.

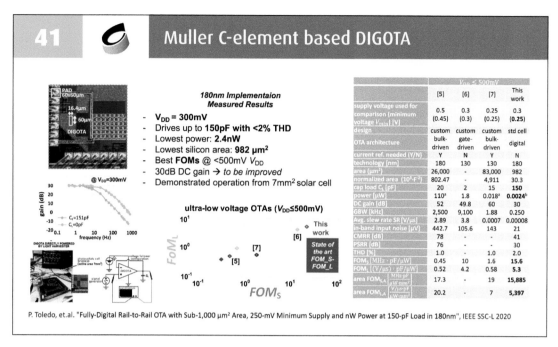

41 — Muller C-element based DIGOTA

180nm Implementaion Measured Results

- V_{DD} = 300mV
- Drives up to **150pF** with <2% THD
- Lowest power: **2.4nW**
- Lowest silicon area: **982 μm²**
- Best **FOMs** @ <500mV V_{DD}
- 30dB DC gain → to be improved
- Demonstrated operation from 7mm² solar cell

ultra-low voltage OTAs ($V_{DD} \leq 500$mV)

	[5]	[6]	[7]	This work
				$V_{DD} < 500mV$
supply voltage used for comparison (minimum voltage V_{min}) [V]	0.5 (0.45)	0.3 (0.3)	0.25 (0.25)	0.3 (0.25)
design	custom	custom	custom	std cell
OTA architecture	bulk-driven	gate-driven	bulk-driven	digital
current ref. needed (Y/N)	Y	N	Y	N
technology [nm]	180	130	130	180
area [μm²]	26,000	-	83,000	982
normalized area ($10^2 \cdot F^2$)	802.47	-	4,911	30.3
cap load C_L [pF]	20	2	15	150
power [μW]	110[a]	1.8	0.018[a]	0.0024[b]
DC gain [dB]	52	49.8	60	30
GBW [kHz]	2,500	9,100	1.88	0.250
Avg. slew rate SR [V/μs]	2.89	3.8	0.0007	0.00008
in-band input noise [μV]	442.7	105.6	143	21
CMRR [dB]	78	-	-	41
PSRR [dB]	76	-	-	30
THD [%]	1.0	-	1.0	2.0
FOM_S [MHz·pF/μW]	0.45	10	1.6	15.6
FOM_L [(V/μs)·pF/μW]	0.52	-	0.58	5.3
area $FOM_{S,A}$ [MHz·pF / μW·mm²]	17.3	-	19	15,885
area $FOM_{L,A}$ [(V/μs)·pF / μW·mm²]	20.2	-	7	5,397

P. Toledo, et.al. "Fully-Digital Rail-to-Rail OTA with Sub-1,000 μm² Area, 250-mV Minimum Supply and nW Power at 150-pF Load in 180nm", IEEE SSC-L 2020

The circuit has also been demonstrated on silicon in 180nm, it occupies an area of less that 1,000 um^2, operates to VDD<300mV driving up to 150pF with a power in the nW range which has been proven to be suitable for direct operation from an energy harvester (a photodiode with less than 7mm^2 area) and achieve state of the art FOMs among OTAs working below 500mV. This confirms that this circuit is suitable to address the challenges of IoT applications.

42 — Muller C-element based DIGOTA

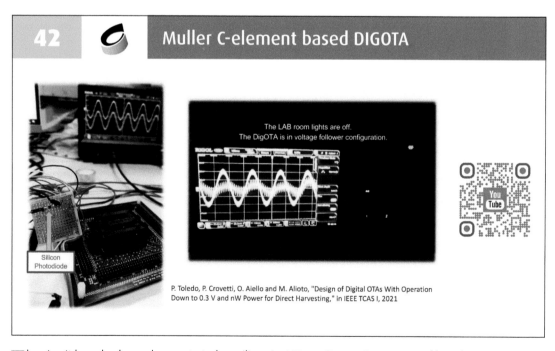

The LAB room lights are off.
The DigOTA is in voltage follower configuration.

P. Toledo, P. Crovetti, O. Aiello and M. Alioto, "Design of Digital OTAs With Operation Down to 0.3 V and nW Power for Direct Harvesting," in IEEE TCAS I, 2021

The circuit has also been demonstrated on silicon in 180nm, it occupies an area of less that 1,000 um^2, operates to VDD<300mV driving up to 150pF with a power in the nW range which has been proven to be suitable for direct operation from an energy harvester (a photodiode with less than 7mm^2 area) and achieve state of the art FOMs among OTAs working below 500mV. This confirms that this circuit is suitable to address the challenges of IoT applications.\

43 Summary

- **Introduction**
  - IoT
  - Analog Design Challenges
  - Highly-Digital Design trend for analog/RF blocks
- **Digital-Based Analog Differential Circuit**
- **Virtual Voltage Reference**
- **Bitstream D/A Conversion Techniques**
  - Relaxation D/A Conversion
- **Conclusion**

44 Voltage Reference

- Accurate voltages and/or currents are needed in:
  - A/D and D/A conversion
  - Power electronics
  - ...
- Semiconductor devices are strongly affected by process, voltage and temperature (PVT) variations
- **Voltage and current references** are cornerstones in analog electronics.

Can we translate it into digital?

Kuijk's Bandgap Reference

K. E. Kuijk, "A precision reference voltage source," IEEE JSSC, 1973.

As we know, reference voltages and currents which are insensitive to process, supply and temperature variations are essential in ICs and the circuits like the Kuijk bandgap reported here, which generate these references are cornerstones in analog electronics. So the question is: can we translate reference circuits into digital?

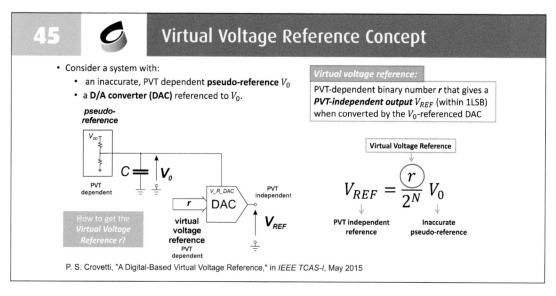

45 Virtual Voltage Reference Concept

- Consider a system with:
  - an inaccurate, PVT dependent **pseudo-reference** V_0
  - a **D/A converter (DAC)** referenced to V_0.

Virtual voltage reference:

PVT-dependent binary number **r** that gives a **PVT-independent output** V_{REF} (within 1LSB) when converted by the V_0-referenced DAC

Virtual Voltage Reference

$$V_{REF} = \frac{r}{2^N} V_0$$

P. S. Crovetti, "A Digital-Based Virtual Voltage Reference," in *IEEE TCAS-I*, May 2015

For this purpose, let's start introducing the virtual voltage reference concept. Let's suppose to have a digital-to-analog converter, which can be implemented in practice by a fully digital circuit (as I will discuss in the final part of my talk), and which is referenced to a pseudo-reference voltage, i.e. an inaccurate and PVT-dependent voltage can be obtained for instance by a voltage divider from an inaccurate supply voltage. With reference to this circuit, the virtual voltage reference is defined as the binary number r, which depends in general on process, voltage and temperature, that, if converted by the DAC referenced to the inaccurate pseudo-reference V0, gives an output voltage Vref which is independent of PVT within 1LSB at the D/A converter resolution. It is quite clear that, if we know this virtual voltage reference r, it is straightforward to obtain a physical voltage reference Vr, but in some cases it is even not essential!

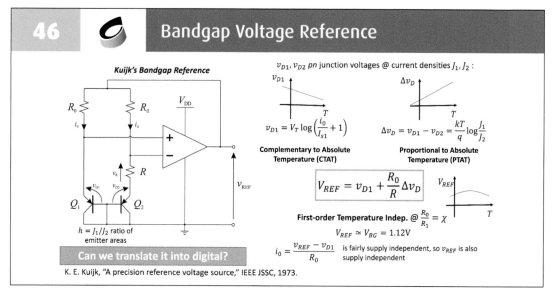

46 Bandgap Voltage Reference

Kuijk's Bandgap Reference

v_{D1}, v_{D2} pn junction voltages @ current densities J_1, J_2 :

$$v_{D1} = V_T \log\left(\frac{i_0}{I_{s1}} + 1\right)$$

Complementary to Absolute Temperature (CTAT)

$$\Delta v_D = v_{D1} - v_{D2} = \frac{kT}{q}\log\frac{J_1}{J_2}$$

Proportional to Absolute Temperature (PTAT)

$$V_{REF} = v_{D1} + \frac{R_0}{R}\Delta v_D$$

First-order Temperature Indep. @ $\frac{R_0}{R_1} = \chi$

$$V_{REF} \simeq V_{BG} = 1.12V$$

$h = J_1//J_2$ ratio of emitter areas

Can we translate it into digital?

$$i_0 = \frac{V_{REF} - v_{D1}}{R_0}$$ is fairly supply independent, so v_{REF} is also supply independent

K. E. Kuijk, "A precision reference voltage source," IEEE JSSC, 1973.

Now the point is: how to calculate in practice the virtual voltage reference r? And the answer is: by translating into digital the operation of a traditional analog voltage reference, like Kujik's circuit! If we focus on the generation of a temperature independent voltage, for instance, we know that in Kujik's circuit, a first-order temperature independent reference voltage is obtained by summing the forward voltage of a pn junction, which is known to be complementary to the absolute temperature (CTAT), and the difference of the forward voltages of two pn junctions biased at different current densities, which is proportional to the absolute temperature (PTAT). Now, can we translate this temperature compensation mechanism into digital?

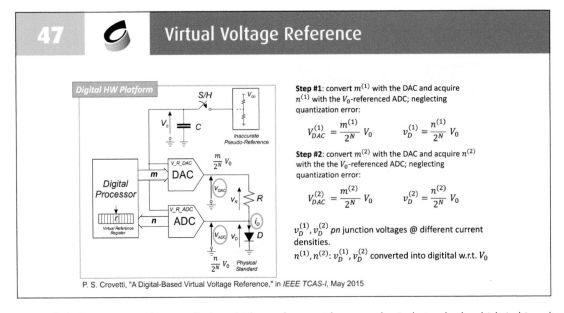

47 **Virtual Voltage Reference**

Step #1: convert $m^{(1)}$ with the DAC and acquire $n^{(1)}$ with the V_0-referenced ADC; neglecting quantization error:

$$V_{DAC}^{(1)} = \frac{m^{(1)}}{2^N} V_0 \qquad v_D^{(1)} = \frac{n^{(1)}}{2^N} V_0$$

Step #2: convert $m^{(2)}$ with the DAC and acquire $n^{(2)}$ with the the V_0-referenced ADC; neglecting quantization error:

$$V_{DAC}^{(2)} = \frac{m^{(2)}}{2^N} V_0 \qquad v_D^{(2)} = \frac{n^{(2)}}{2^N} V_0$$

$v_D^{(1)}, v_D^{(2)}$ pn junction voltages @ different current densities.

$n^{(1)}, n^{(2)}: v_D^{(1)}, v_D^{(2)}$ converted into digitital w.r.t. V_0

P. S. Crovetti, "A Digital-Based Virtual Voltage Reference," in *IEEE TCAS-I*, May 2015

Well, let's suppose to have a diode, which can be consider as a physical standard, which is biased through a resistor R by the output voltage of a DAC, referenced to an inaccurate pseudo-reference voltage V0, and let's suppose the diode forward voltage is acquired and converted into digital by an ADC, referenced to the same pseudo-reference, and made available to a digital processor. If we convert two different values m1 and m2 into analog by the DAC, we can bias the diode at two different current densities and we can get from the analog to digital converter the binary numbers n1 and n2 corresponding to the two diode forward voltages v_D(1) and v_D(2) converted into digital with respect to the pseudo-reference V0 (neglecting quantization error).

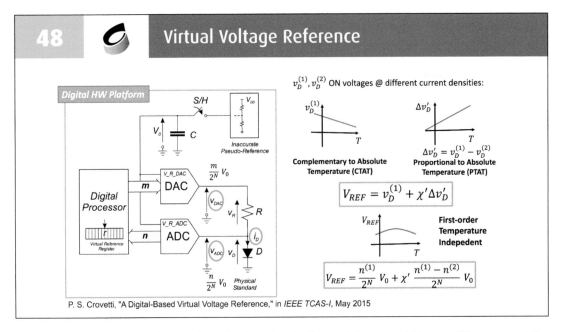

48 **Virtual Voltage Reference**

$v_D^{(1)}, v_D^{(2)}$ ON voltages @ different current densities:

Complementary to Absolute Temperature (CTAT)

Proportional to Absolute Temperature (PTAT)

$$\Delta v_D' = v_D^{(1)} - v_D^{(2)}$$

$$\boxed{V_{REF} = v_D^{(1)} + \chi' \Delta v_D'}$$

First-order Temperature Indepedent

$$\boxed{V_{REF} = \frac{n^{(1)}}{2^N} V_0 + \chi' \frac{n^{(1)} - n^{(2)}}{2^N} V_0}$$

P. S. Crovetti, "A Digital-Based Virtual Voltage Reference," in *IEEE TCAS-I*, May 2015

If we consider the diode voltages vd1 and vd2 under the first and the second bias conditions, we get the vd1 is CTAT while their difference is PTAT. So, if we sum these voltages with an appropriate weight, we can get a first-order temperature independent voltage as in Kuijk circuit. By the way, we do not have these physical voltages in our circuit, we just know the results of their conversion into digital. But if we replace vd1 and delta vd' with their expressions in terms of the ADC acquired valued n1 and n2, we can get this expression

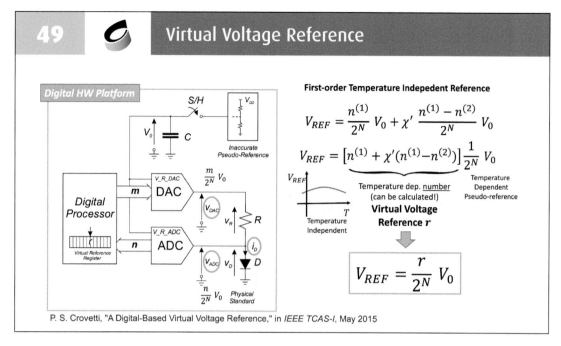

49 Virtual Voltage Reference

First-order Temperature Indepedent Reference

$$V_{REF} = \frac{n^{(1)}}{2^N} V_0 + \chi' \frac{n^{(1)} - n^{(2)}}{2^N} V_0$$

$$V_{REF} = [n^{(1)} + \chi'(n^{(1)} - n^{(2)})] \frac{1}{2^N} V_0$$

Temperature dep. number (can be calculated!)

Temperature Dependent Pseudo-reference

Virtual Voltage Reference r

$$V_{REF} = \frac{r}{2^N} V_0$$

P. S. Crovetti, "A Digital-Based Virtual Voltage Reference," in *IEEE TCAS-I*, May 2015

I f we collect V0/2^N, we can observe that this expression is analogous to the definition of the virtual voltage reference! This factor here is a binary number, which can be calculated by the digital processor starting from the data acquired by the ADC and is fully equivalent to the virtual voltage reference! In other words, this quantity, which can be calculated algorithmically inside the digital processor, is our virtual voltage reference!

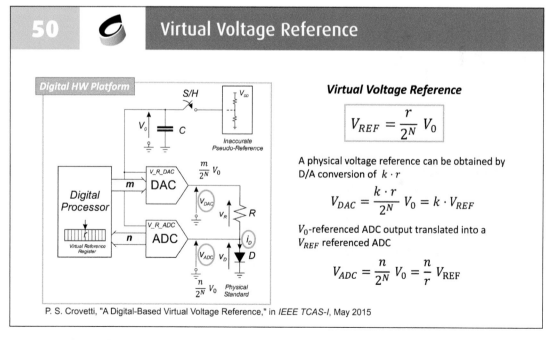

50 Virtual Voltage Reference

Virtual Voltage Reference

$$V_{REF} = \frac{r}{2^N} V_0$$

A physical voltage reference can be obtained by D/A conversion of $k \cdot r$

$$V_{DAC} = \frac{k \cdot r}{2^N} V_0 = k \cdot V_{REF}$$

V_0-referenced ADC output translated into a V_{REF} referenced ADC

$$V_{ADC} = \frac{n}{2^N} V_0 = \frac{n}{r} V_{REF}$$

P. S. Crovetti, "A Digital-Based Virtual Voltage Reference," in *IEEE TCAS-I*, May 2015

O nce we know the virtual voltage reference, we can get a physical reference just by converting r into analog by the DAC. Moreover, if we acquire an analog voltage with the ADC referenced to V0, we can express V0 in terms of Vref based on the definition of the virtual voltage, we can translate the acquisition made by the ADC referenced to the pseudo-reference into the acquisition made of a virtual ADC referenced to the PVT-independent reference VREF, even if VREF does not exist as a physical voltage.

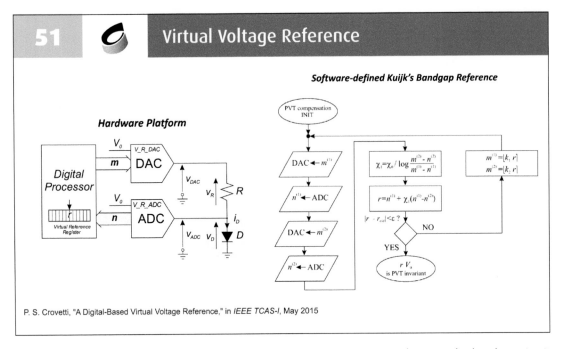

51 Virtual Voltage Reference

Software-defined Kuijk's Bandgap Reference

Hardware Platform

P. S. Crovetti, "A Digital-Based Virtual Voltage Reference," in *IEEE TCAS-I*, May 2015

By this approach, we can implement the temperature compensation mechanism of a bandgap circuit by a software procedure. By including other steps in this procedure, it is also possible to compensate power supply voltage and pseudo-reference variations as described in the referenced paper.

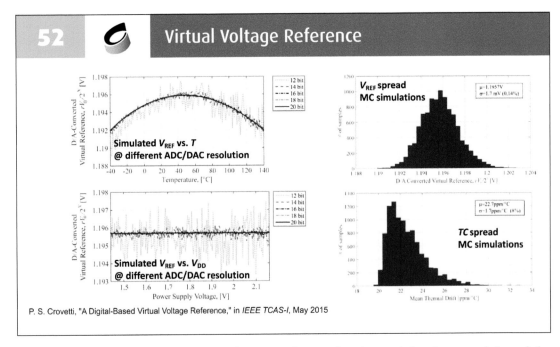

52 Virtual Voltage Reference

Simulated V_{REF} vs. T @ different ADC/DAC resolution

V_{REF} spread MC simulations

Simulated V_{REF} vs. V_{DD} @ different ADC/DAC resolution

TC spread MC simulations

P. S. Crovetti, "A Digital-Based Virtual Voltage Reference," in *IEEE TCAS-I*, May 2015

This approach has been tested by simulations, and it can be observed that for a resolution of the ADC/DAC exceeding 12-13 bits, the thermal drift of the virtual referenced, re-converted into analog, is analogous to that of a Kuijk bandgap reference, and also the residual power supply dependence is very low. Moreover, since just a single diode is employed in this circuit, it is particularly robust against process variations and mismatch (the std dev of the reference voltage over process variations is just 0.14% for a 10um x10um diode). Also the spread of the thermal drift is quite low.

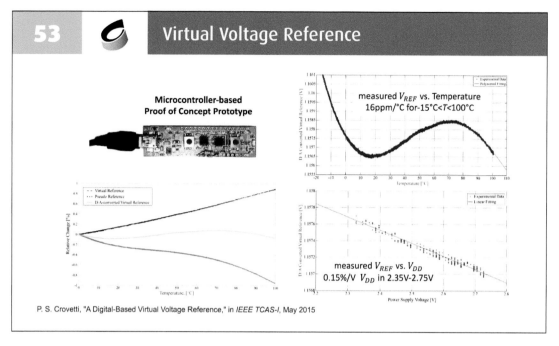

53 — Virtual Voltage Reference

Microcontroller-based Proof of Concept Prototype

measured V_{REF} vs. Temperature
16ppm/°C for -15°C<T<100°C

measured V_{REF} vs. V_{DD}
0.15%/V V_{DD} in 2.35V-2.75V

P. S. Crovetti, "A Digital-Based Virtual Voltage Reference," in *IEEE TCAS-I*, May 2015

Measurements on a microntroller-based virtual voltage reference prototype show a thermal drift of 16ppm/°C in a temperature range from -15°C to 100°C, and this plot which shows the measured results, it can be observed that the supply voltage of the microntroller, used as a pseudo-reference, shows a thermal drift of more than 1%, which is fully compensated by the opposite drift of the virtual voltage reference, so that to obtain a fairly temperature independent reference voltage.

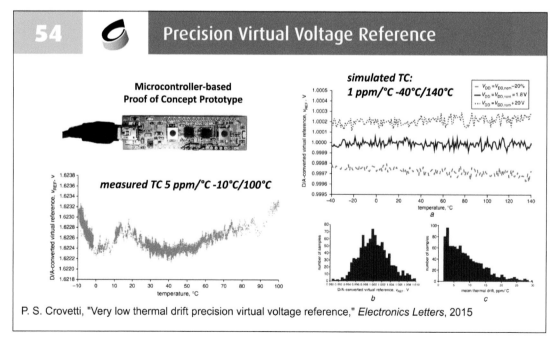

54 — Precision Virtual Voltage Reference

Microcontroller-based Proof of Concept Prototype

simulated TC:
1 ppm/°C -40°C/140°C

$V_{DD} = V_{DD,nom} - 20\%$
$V_{DD} = V_{DD,nom} = 1.8V$
$V_{DD} = V_{DD,nom} + 20V$

measured TC 5 ppm/°C -10°C/100°C

a

b

c

P. S. Crovetti, "Very low thermal drift precision virtual voltage reference," *Electronics Letters*, 2015

One of the potential advantages of this solution is that since the processing part is moved into digital, it is possible to implement complex, high order temperature-compensation strategies which would not be suitable to an analog implementation. This approach has been adopted in a second precision virtual voltage reference which achieves a measured temperature coefficient which is 5ppm/°C.

55 Independent silicon verification

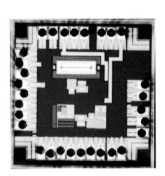

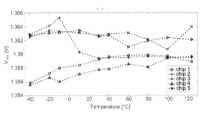

Simpler 'virtual reference' concept (independently reported later) adopted in Algorithmic Resistance and Power measurement IC developed at TU-Delft

Silicon validation: TC 18ppm/°C in line with the microcontroller prototype

Temperature range	−40°C – 125°C
Temperature inaccuracy	±0.25°C (min-max)
Temperature dependence of voltage reference	18 ppm/°C
Resistance inaccuracy	±0.55 Ω[†]
Power inaccuracy	±0.8%[†]

[†]after a single temperature calibration for the on-chip reference resistor, 5 samples.

Z. Cai *et al.*, "A CMOS Readout Circuit for Resistive Transducers Based on Algorithmic Resistance and Power Measurement," in *IEEE Sensors Journ., 2017*

This concept has also been verified on silicon (even if not by my research group). It has been implemented two years after the first publication and it can be observed that the silicon implementation in 180nm achieves a temperature coefficient of 18ppm/°C which is fully in line with the results obtained by the microcontroller based prototype. This confirms that the digital based approach can be reliably tested by standard digital HW (microcontrollers, FPGA) with results which are compatible with the silicon implementation.

56 Summary

- **Introduction**
  - IoT
  - Analog Design Challenges
  - Highly-Digital Design trend for analog/RF blocks
- **Digital-Based Analog Differential Circuit**
- **Virtual Voltage Reference**
- **Bitstream D/A Conversion Techniques**
  - **Relaxation D/A Conversion**
- **Conclusion**

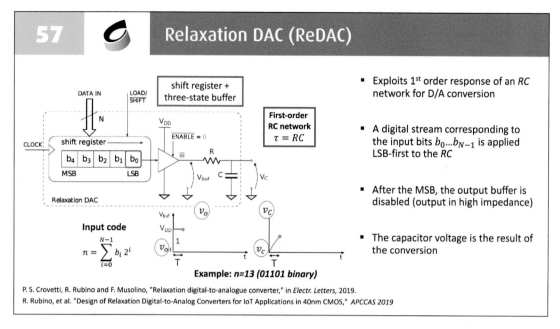

P. S. Crovetti, R. Rubino and F. Musolino, "Relaxation digital-to-analogue converter," in *Electr. Letters, 2019.*

R. Rubino, et al. "Design of Relaxation Digital-to-Analog Converters for IoT Applications in 40nm CMOS," *APCCAS 2019*

A relaxation DAC exploits the time response of a first-order RC network to perform digital to analog conversion. It includes a shift register, loaded with the digital input to be converted and that drives the RC through a three-state buffer, with a digital stream corresponding to the bits of the input code, starting from the LSB. After the conversion of the last bit, the voltage across the capacitor, which is kept constant by disabling the three-buffer, for a particular value of the clock period, can be shown to be proportional to the digital input to be converted.

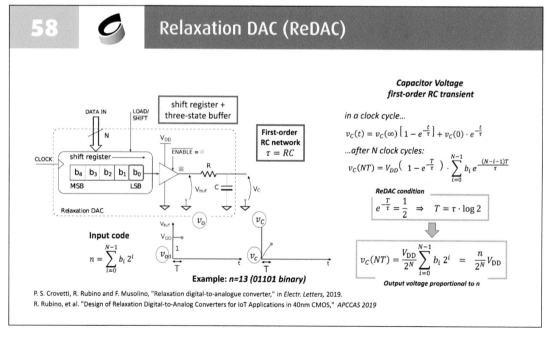

P. S. Crovetti, R. Rubino and F. Musolino, "Relaxation digital-to-analogue converter," in *Electr. Letters, 2019.*

R. Rubino, et al. "Design of Relaxation Digital-to-Analog Converters for IoT Applications in 40nm CMOS," *APCCAS 2019*

This property can be easily shown studying the evolution of the capacitor voltage in a clock period by the general expression of the transient in a first-order linear network. By iterating the expression for N clock cycles, considering the values of the input bits, we get the final capacitor voltage and we can observe that, if we impose in this expression a relation between the clock period and the time constant tau of the RC network, i.e. T=tau log 2, we get the expression of the capacitor voltage is VDD/2^N times the value of the value of the binary input n, as expected in a digital to analog converter.

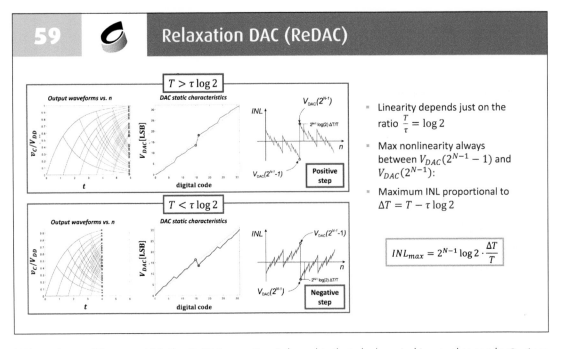

The only condition on which the ReDAC operation is based is that clock period is equal to tau log2, since we can observe that if T>tau log2, we have a positive step between two mid-range values, whereas if T<tau log2, we have a negative step between the same converted values. By checking the difference between these values and by tuning the clock period so that to enforce that this difference is zero, we can enforce actually the correct ReDAC operation of the circuit.

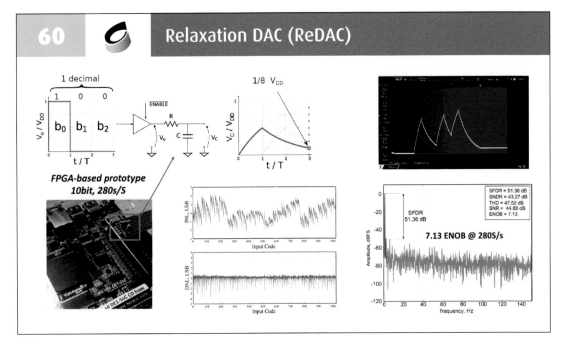

The ReDAC concept has been experimentally demonstrated on FPGA, and 7ENOB resolution at 280 s/S has been demonstrated as a proof of concept. A possible implementation of this calibration strategy has been shown in a paper presented in this conference, in the Data Converters III session, and invite you to have a look to the paper and to the video of the presentation for more details.

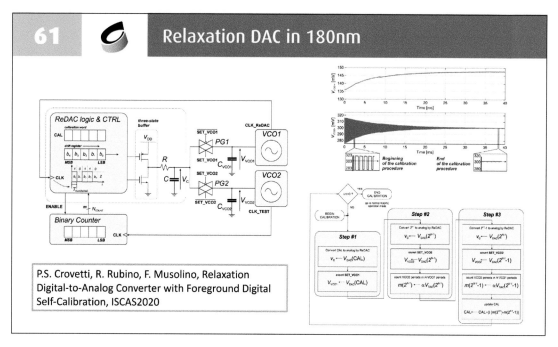

61 Relaxation DAC in 180nm

P.S. Crovetti, R. Rubino, F. Musolino, Relaxation Digital-to-Analog Converter with Foreground Digital Self-Calibration, ISCAS2020

The ReDAC concept has been experimentally demonstrated on FPGA, and 7ENOB resolution at 280 s/S has been demonstrated as a proof of concept. A possible implementation of this calibration strategy has been shown in a paper presented in this conference, in the Data Converters III session, and invite you to have a look to the paper and to the video of the presentation for more details.

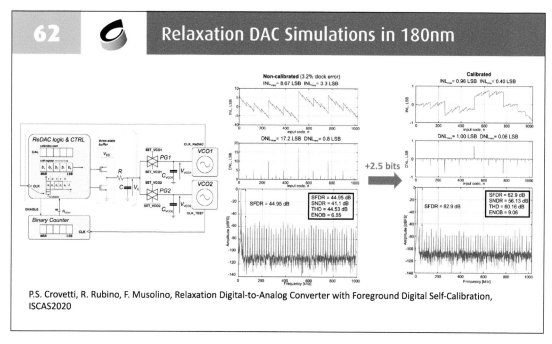

62 Relaxation DAC Simulations in 180nm

P.S. Crovetti, R. Rubino, F. Musolino, Relaxation Digital-to-Analog Converter with Foreground Digital Self-Calibration, ISCAS2020

Based on simulations of the same circuit in 40nm, very interesting results have been obtained: a 10bit, 2MS/s ReDAC has been implemented and, based on post layout simulations, the energy per conversion is very low: 0.73pJ/conv including the digital part, and just 210fJ for the output stage, yielding a very competitive FOM of about 1fJ/conv-step.

63 FPGA-Based ReDAC with Parasitics Error Suppresion

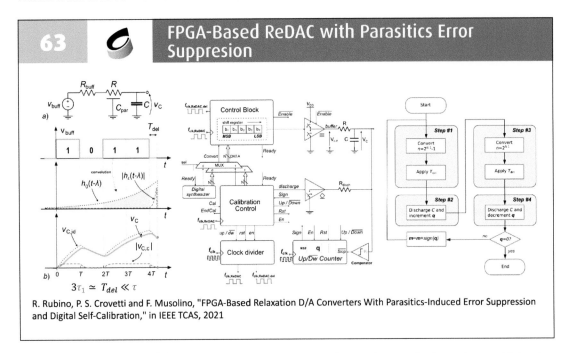

R. Rubino, P. S. Crovetti and F. Musolino, "FPGA-Based Relaxation D/A Converters With Parasitics-Induced Error Suppression and Digital Self-Calibration," in IEEE TCAS, 2021

Based on simulations of the same circuit in 40nm, very interesting results have been obtained: a 10bit, 2MS/s ReDAC has been implemented and, based on post layout simulations, the energy per conversion is very low: 0.73pJ/conv including the digital part, and just 210fJ for the output stage, yielding a very competitive FOM of about 1fJ/conv-step.

64 FPGA-Based ReDAC with Parasitics Error Suppresion

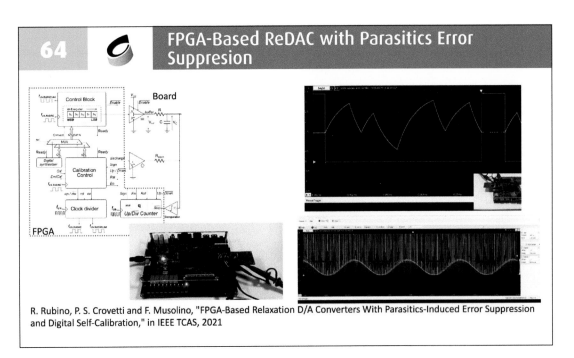

R. Rubino, P. S. Crovetti and F. Musolino, "FPGA-Based Relaxation D/A Converters With Parasitics-Induced Error Suppression and Digital Self-Calibration," in IEEE TCAS, 2021

Based on simulations of the same circuit in 40nm, very interesting results have been obtained: a 10bit, 2MS/s ReDAC has been implemented and, based on post layout simulations, the energy per conversion is very low: 0.73pJ/conv including the digital part, and just 210fJ for the output stage, yielding a very competitive FOM of about 1fJ/conv-step.

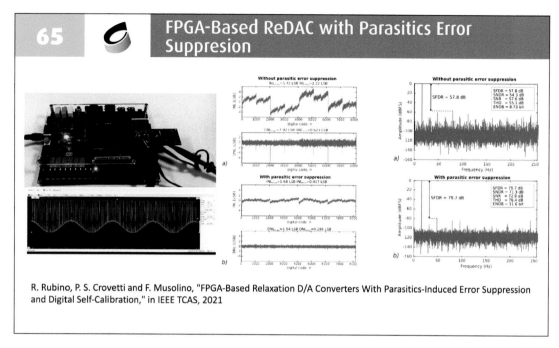

65 **FPGA-Based ReDAC with Parasitics Error Suppresion**

R. Rubino, P. S. Crovetti and F. Musolino, "FPGA-Based Relaxation D/A Converters With Parasitics-Induced Error Suppression and Digital Self-Calibration," in IEEE TCAS, 2021

Based on simulations of the same circuit in 40nm, very interesting results have been obtained: a 10bit, 2MS/s ReDAC has been implemented and, based on post layout simulations, the energy per conversion is very low: 0.73pJ/conv including the digital part, and just 210fJ for the output stage, yielding a very competitive FOM of about 1fJ/conv-step.

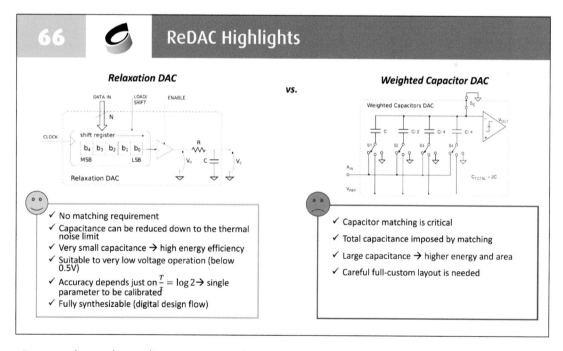

66 **ReDAC Highlights**

Relaxation DAC **vs.** *Weighted Capacitor DAC*

✓ No matching requirement
✓ Capacitance can be reduced down to the thermal noise limit
✓ Very small capacitance → high energy efficiency
✓ Suitable to very low voltage operation (below 0.5V)
✓ Accuracy depends just on $\frac{T}{\tau} = \log 2$ → single parameter to be calibrated
✓ Fully synthesizable (digital design flow)

✓ Capacitor matching is critical
✓ Total capacitance imposed by matching
✓ Large capacitance → higher energy and area
✓ Careful full-custom layout is needed

Compared to other D/A conversion techniques, this technique does not impose any matching requirements, so that the output capacitance can be reduced down to the limit dictated by thermal noise. Very small capacitance also mean very high energy efficiency. Moreover, since the accuracy of the conversion depends just on the relation between the time constant and the clock period, it is possible to calibrate this circuit by enforcing just a single condition, and both the DAC and the calibration network can be fully synthesized in a digital design flow. These features make this circuit convenient and also competitive with weighted capacitors digital to analog converters, in IoT applications.

67   Summary

- Introduction
  - IoT
  - Analog Design Challenges
  - Highly-Digital Design trend for analog/RF blocks
- Digital-Based Analog Differential Circuit
- Virtual Voltage Reference
- Bitstream D/A Conversion Techniques
  - Relaxation D/A Conversion
- Conclusion

68   Conclusions

- **Analog processing by digital circuits is possible**
- Analog interfaces made by *true* digital circuits address the requirements of the IoT:
  - Scaling friendly, matching insensitive and robust against PVT variations
  - Ultra Low Voltage (ULV), Ultra Low Power (ULP), energy efficient operation
  - Reconfigurability and fast testing/prototyping (e.g. FPGA-based)
  - Fully automated design flow
  - Portability across technology nodes
- No *fundamental* performance limitation in digital-based design

I hope I have convinced you with several examples that analog processing by digital circuits is feasible and even analog building blocks like operational transconductance amplifiers and voltage references can be conveniently translated into digital so that to be implemented by true digital circuits. The circuits which can be obtained by this approach are suitable to address the challenges of analog interfaces in IoT applications, since they are scaling friendly, they are suitable to operation at ultra-low supply voltage and ultra-low power consumption; they are reconfigurable, they can be designed by a fully automated design flow and are easily portably across technology nodes, thus reducing cost, time to market and design effort in a significant way. For what concerns performance, even if the prototypes that have been discussed in this presentation show some limitations compared to the best analog solutions, I would like to stress that there is not intrinsic performance limitation in this approach and, unlike in analog solutions and the circuits

69 Conclusions

- **Analog processing by digital circuits is possible**
- The boundary between analog and digital is actually very faint
  - Digital design concepts can be borrowed in analog design
  - Analog design concepts could be possibly applied to address the challenges of digital design (big data, machine learning,...)?
- An unified, cross-domain analog/digital approach is needed to address the challenges of future information processing.

are going to take advantage of scaling, so improvements can be expected by refining the design. From a more general perspective it can be concluded that the boundary between analog and digital is very faint: as we have seen, digital design concepts can be borrowed and conveniently applied in analog design. But, as an analog designer, I also believe that the opposite is true: analog design concepts can be possibly applied to address the challenges of digital information processing in big data, machine learning, etc.. In conclusion I do not believe that it makes any more sense to make a strict distinction between analog and digital, but a cross-domain analog and digital approach at the same time is needed in the future of information processing.

Design and Development of Depleted Monolithic Active Pixel Sensors for High-Radiation Applications

Dr. Konstantinos Moustakas

ASIC Design, Paul Scherrer Insitut (PSI),
Switzerland

INTRODUCTION

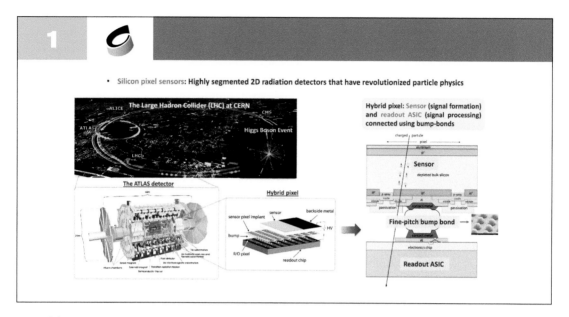

- **Silicon pixel sensors:** Highly segmented 2D radiation detectors that have revolutionized particle physics

Pixel detectors are an essential part of experimental high-energy physics instrumentation. They constitute tracking detectors that image the trajectories of charged particles and are highly segmented and fast i.e. able to capture millions of events (images) per second. The established hybrid pixel technology consists of two parts, the sensor where a signal of electron hole pairs (e/h) is generated by ionization from a traversing particle and the readout electronics that process, digitize, store and transmit the hit data. The two parts are connected together via fine-pitch bump bonds.

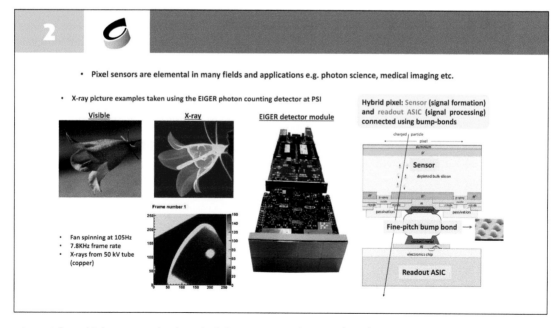

- **Pixel sensors are elemental in many fields and applications e.g. photon science, medical imaging etc.**

- **X-ray picture examples taken using the EIGER photon counting detector at PSI**

Apart from high-energy physics, pixel detectors are elemental in photon science where they are used as X-ray imaging devices. Being used in synchrotron and free electron laser (FEL) facilities around the world, they enable cutting-edge research in many fields such as material science and biology (protein structure and folding). Additionally, the constitute a critical component of many medical imaging devices.

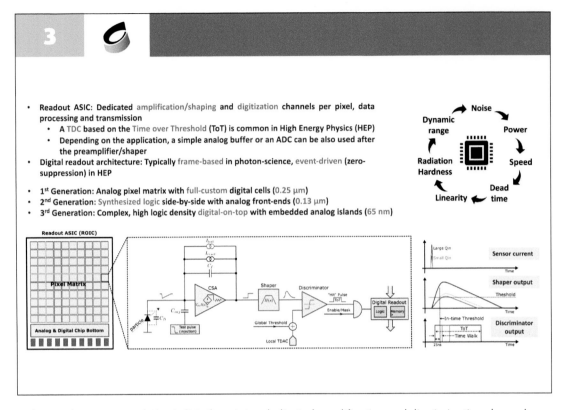

- Readout ASIC: Dedicated amplification/shaping and digitization channels per pixel, data processing and transmission
  - A TDC based on the Time over Threshold (ToT) is common in High Energy Physics (HEP)
  - Depending on the application, a simple analog buffer or an ADC can be also used after the preamplifier/shaper
- Digital readout architecture: Typically frame-based in photon-science, event-driven (zero-suppression) in HEP

- 1st Generation: Analog pixel matrix with full-custom digital cells (0.25 μm)
- 2nd Generation: Synthesized logic side-by-side with analog front-ends (0.13 μm)
- 3rd Generation: Complex, high logic density digital-on-top with embedded analog islands (65 nm)

The Readout Integrated Circuit (ROIC) contains dedicated amplification and discrimination channels per pixel (front-end) that operate in parallel followed by parallel processing of the output signals which are most commonly digitized in the pixel. As technology evolves and pixel detectors are required to cope with ever increasing particle fluence, ROIC's have become complex and high logic density digital chips with embedded analog front-ends in a sea of digital gates (digital-on-top design flow). A particle impinging at the sensor diode, induces a short current signal at the pre-amplifier input. The pre-amplifier is usually realized by an active integrator and generates a voltage proportional to the signal charge. A shaper, which is essentially a band-pass filter with properly tuned time constant, is usually employed to optimize the signal shape and bandwidth in order to reduce noise and pile-up. The discriminator compares the pre-amplifier output signal to a reference voltage. Whenever the signal amplitude exceeds the reference voltage , a digital pulse with width (Time-over-Threshold) proportional to the amount of collected charge is produced. The digital readout architecture is tasked with transferring data from the pixels to the ROIC output. While in photon science full images are typically captured using a frame-based readout, high-energy physics applications employ a zero-suppression readout which only processes pixels with amplitudes above the threshold set by the discriminator.

4

- **Charge Sensitive Amplifier (CSA) → capacitive feedback loop (integrator)**
  - Preferable in applications with small input signal / high C_D
  - "Forces" the sensor current through the Miller capacitor (C_M)

$$C_M = C_f(1 + A_0) \qquad V_{out} = -\frac{Q}{C_f} \frac{1}{1 + \frac{1}{A_0} + \frac{C_{in}}{A_0 C_f}} \cong -\frac{Q}{C_f}$$

- **In reality the amplifier band-limited. The CSA response time is (single pole amplifier):**

$$\tau_{CSA} \cong \frac{C_D}{C_f \cdot GBW} \propto \frac{C_D}{C_f \cdot g_m}$$

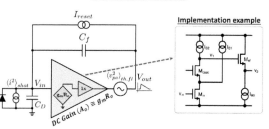

Implementation example

- **The input shot noise current flows through C_f**
- **The pre-amp output noise (flicker, thermal) is fed back to the input by the capacitive divider $\frac{C_f}{C_D}$**
- **The input transistor noise is usually dominant:**

$$\frac{d\langle v_{CSA}^2 \rangle}{d\omega} = \underbrace{\frac{eI_0}{\pi \omega^2 C_f^2}}_{\text{shot}} + \underbrace{\frac{K_f}{C_{ox}^2 WL}\frac{C_D^2}{C_f^2}\frac{1}{\omega}}_{\text{flicker}} + \underbrace{\frac{4\,kT}{3\,g_m}\frac{C_D^2}{C_f^2}}_{\text{thermal}}$$

- **Assuming a shaper (CR^1-RC^1) in the chain, the Equivalent Noise Charge (ENC) is:**

$$ENC^2 \propto \underbrace{eI_0\tau_{sh}}_{\text{shot}} + \underbrace{\frac{K_f}{C_{ox}^2 WL}C_D^2}_{\text{flicker}} + \underbrace{\frac{4\,kT}{3}\frac{C_D^2}{g_m\tau_{sh}}}_{\text{thermal}}$$

- **An optimal shaping time exists!**

The pre-amplifier of a ROIC front-end chain is commonly implemented using a capacitive feedback loop and is referred to as a Charge Sensitive Amplifier (CSA). Its operation principle is based on the integration of the input charge Q on the feedback capacitor Cf with results in an output voltage amplitude approximately equal to Q/Cf. In order to "force" the input current through the feedback capacitor instead of the usually much larger sensor capacitance, a low impedance is generated at the pre-amplifier input by multiplying the feedback capacitance by the amplifier open-loop gain according to the miller effect. In reality the pre-amplifier frequency response (bandwidth) has to be taken into account. At low frequencies, the CSA input impedance is purely capacitive while at high frequencies the amplifier behaves like a resistor (assuming a single dominant pole). Therefore, the CSA response time is proportional to the sensor capacitance (Cd) and inversely proportional to the amplifier gain-bandwidth product (GBW). Noise is one of the most crucial performance metrics of the pre-amplifier. It is commonly expressed as the ratio of the output rms noise voltage to the output voltage amplitude of a signal of 1 electron which is referred to as the Equivalent Noise Charge (ENC). The main noise components are the shot noise generated by the sensor diode and the thermal and flicker noise generated by the pre-amplifier MOSFET transistors. Usually, the input transistor noise is dominant and as a result it needs to be sized carefully. Since the ENC is proportional to the sensor capacitance (Cd), the latter is one of the most crucial parameters which influence the analog performance of a ROIC readout chain. The shot noise component is proportional, while the thermal noise component is inversely proportional to the shaping time. Therefore, an optimum time constant exists for which the total noise is minimal.

Radiation Effects		
Bulk damage (NIEL)	**Surface damage (TID)**	**Single event effects (SEE)**

Bulk damage (NIEL)

- Traversing particles can dislocate atoms → vacancies and interstitials
- Non-Ionizing Energy Loss (NIEL) process
- Described in terms of neutron equivalent fluence (n_{eq}/cm^2)
- Changes of the substrate doping
- Increases the sensor leakage current (shot noise)
- Higher charge trapping probability → lower charge collection efficiency

Surface damage (TID)

- Charge carriers trapped in oxide structures → accumulation of positive charge
- Proportional to the Total Ionizing Dose (TID)
- STI and spacers (thick oxide) are affected
- Short, narrow channel effects
  - NMOS → leakage
  - PMOS → "slower"
- Parasitic NMOS (STI)

Single event effects (SEE)

- Free carriers generated by ionization collected at a sensitive circuit node

- Single Event Transient (SET) → "glitch"
- Single Event Upset (SEU) → bit flipping
- Single Event Latch-up (SEL)
- Single Event Gate Rupture (SEGT)

Radiation hardening

- Triple Modular Redundancy (TMR)
- Specialized logic (e.g. DICE latch)
- Error tolerant encoding (e.g. Hamming)

Radiation hardening

- Large devices
- Enclosed layout transistors (ELT)
- p+ guard rings

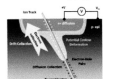

Pixel detector and CMOS ROIC characteristics are affected by radiation damage that occurs during their lifetime. Therefore, these effects should be taken into account and necessary measures need to be employed in order to improve the radiation tolerance (radiation-hardening) of a ROIC according to the application. Radiation damage effects can be divided in three main categories. The first category is damage to the silicon crystal bulk induced by traversing massive particles that create vacancies and interstitials in the lattice. These lead to changes in the substrate doping, increased leakage current (and shot noise) and most importantly higher probability of charge trapping within the sensor crystal which reduces the input signal charge. The second category is surface damage due to positive charge accumulation in silicon oxides that is proportional to the Total Ionizing Dose (TID). Although the MOSFET gate oxide is crucial, in modern technologies it is very thin (in the order of a few nm). Therefore, the accumulated charge can escape due to quantum tunneling and as a result TID effects are most prominent in thick oxides such as the Silicon Trench Isolation (STI) and MOSFET source/drain spacers. The net result of these effects is parasitic leakage in the case of NMOS transistor and reduced current flow in the case of PMOS transistors. Furthermore, high doses can lead to the activation of parasitic NMOS transistors under the STI. In order to increase radiation tolerance to TID large transistor devices are required which are shielded by p+ guard rings. Furthermore, Enclosed Layout Transistors (ELT) that prevent parasitic leakage paths are usually employed. The third category are Single Event Effects (SEE) due to a large number of free carriers generated by ionization. These carriers can be collected by sensitive nodes of the circuit and induce glitches, bit-flips or even latch-up events. Radiation hardening measures include Triple Modular Redundancy (TMR), specialized logic and error tolerant encoding.

PART I: DMAPS

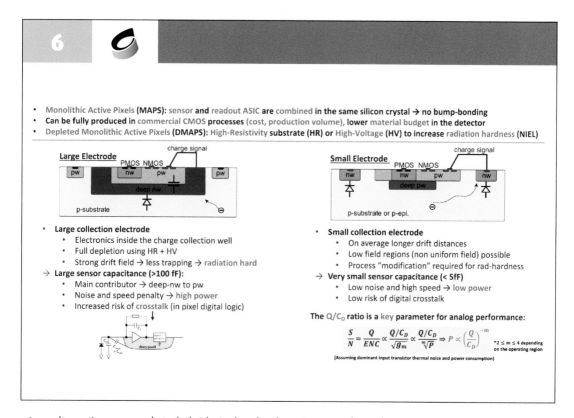

- Monolithic Active Pixels **(MAPS)**: sensor **and** readout ASIC **are combined in the same silicon crystal → no bump-bonding**
- **Can be fully produced in** commercial CMOS **processes** (cost, production volume), **lower** material budget **in the detector**
- Depleted Monolithic Active Pixels **(DMAPS)**: High-Resistivity **substrate (HR)** or High-Voltage **(HV) to increase** radiation hardness **(NIEL)**

An alternative approach to hybrid pixel technology is to combine the sensor and readout electronics in the same silicon crystal and form a monolithic pixel that can be produced using commercial CMOS technologies. Monolithic pixel detectors do not require complex and laborious bump-bonding and flip-chipping and can be produced in large volumes by commercial semiconductor foundries. Therefore, they are cost-effective and ideal for large area detectors. Furthermore, since they constitute a single entity, their thickness can be small (100 – 150 μm), offering low material budget in the detector. Depleted Monolithic Active Pixels (DMAPS) exploit high resistivity substrates, high voltage biasing or a combination of both to achieve full depletion and fast charge collection by drift, rendering them suitable for high-rate high-radiation environments. Multiple nested well technologies employed by DMAPS allow complex CMOS circuitry and high functionality to be integrated in the pixel. Two principal variants of a DMAPS pixel cell arrangement can be implemented depending on the collection electrode geometry. In the case of the large collection electrode implementation concept, the readout electronics are housed inside a large deep n-well which acts are the collecting node. The advantage of this approach is the strong and uniform electric field that results in efficient charge collection, while the disadvantage is the large sensor capacitance (increased noise, power consumption) and the increased risk of crosstalk. On the other hand, in the case of the small collection electrode concept, a small n-well is used as the collecting node and is placed outside the CMOS electronics area. This configuration allows for a very small sensor capacitance (<5fF), which is translated to high analog performance. Furthermore, crosstalk due to digital transients is drastically reduced. However, it is more difficult to achieve high charge collection efficiency compared to the large electrode design due to the on-average longer drift paths (for the same pixel size) and the non-uniform electric field.

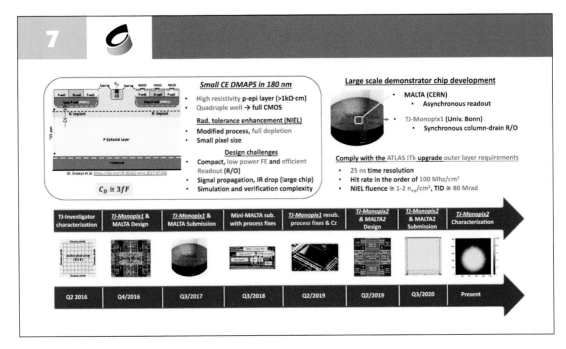

TJ-Monopix is a large scale, small fill-factor demonstrator DMAPS chip with integrated standalone readout architecture and ToT capability. It is part of a DMAPS development line pursued by a collaboration between the University of Bonn and CERN with the goal to combine high analog performance (due to the very small sensor capacitance ~ 3fF) with high radiation tolerance in order to comply with the stringent requirements of the ATLAS inner tracker (ITk) outer layer after the High-Luminosity LHC upgrade. TJ-Monopix is based on a quadruple well 180 nm CMOS imaging process that allows full CMOS in the active pixel matrix and features a high resistivity p-epitaxial substrate. In order to improve radiation tolerance to bulk damage, the process has been modified by CERN and the foundry. A low doping n-type layer is implanted in the epitaxial layer within the pixel matrix below the wells containing circuitry and as a result a large, even, planar-like junction is formed that extends depletion over the full pixel area and enhances charge collection efficiency. TJ-Monopix1 is the first, half scale (1x2 cm2) version and has been extensively measured and characterized. The design was further improved to reduce the pixel size, noise and threshold dispersion and a new, full scale version (2x2 cm2) called TJ-Monopix2 has been developed and is currently being characterized.

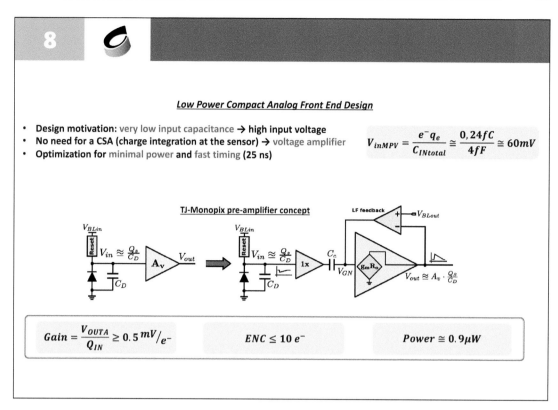

Low Power Compact Analog Front End Design

- **Design motivation:** very low input capacitance → high input voltage
- **No need for a CSA** (charge integration at the sensor) → voltage amplifier
- **Optimization for** minimal power **and** fast timing **(25 ns)**

$$V_{inMPV} = \frac{e^- q_e}{C_{INtotal}} \cong \frac{0,24fC}{4fF} \cong 60mV$$

TJ-Monopix pre-amplifier concept

$$Gain = \frac{V_{OUTA}}{Q_{IN}} \geq 0.5\,mV/e^-$$

$$ENC \leq 10\,e^-$$

$$Power \cong 0.9\mu W$$

In order to exploit the very small sensor capacitance (Cd) and small pixel size of TJ-Monopix, a high performance, low-power, compact analog front-end has been developed as an evolution of the ALPIDE MAPS chip front-end. Due to the large input signal input, as a result of the small Cd, a voltage amplifier is used instead of a CSA. A CSA is not optimal in this case because the input charge is integrated directly at the sensor capacitance which is in the order, if not smaller, compared to a typical CSA feedback capacitance. The input signal is buffered and subsequently amplified by a voltage amplification stage which includes a low frequency feedback to reset-restore the output voltage baseline. The resulting ENC is smaller than 10 electrons with a power consumption of only 0.9 μW.

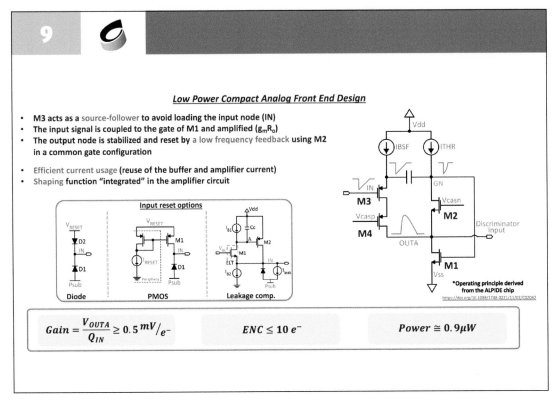

Low Power Compact Analog Front End Design

- M3 acts as a source-follower to avoid loading the input node (IN)
- The input signal is coupled to the gate of M1 and amplified ($g_m R_o$)
- The output node is stabilized and reset by a low frequency feedback using M2 in a common gate configuration

- Efficient current usage (reuse of the buffer and amplifier current)
- Shaping function "integrated" in the amplifier circuit

Input reset options

| Diode | PMOS | Leakage comp. |

Operating principle derived from the ALPIDE chip
https://doi.org/10.1088/1748-0221/11/03/C02042

$$Gain = \frac{V_{OUTA}}{Q_{IN}} \geq 0.5 \, mV/e^- \qquad ENC \leq 10 \, e^- \qquad Power \cong 0.9 \mu W$$

The input signal is buffered by a source follower (M3) and coupled by a MOS capacitor to the gate of M1, which acts as a transconductor. The resulting M1 small signal current is converted to an amplified output voltage at the high impedance node OUTA. M2 is a common gate amplifier, which is part of a low-frequency negative feedback loop that controls the gate of M1 in order to reset the output. Due to its low biasing current, M2 is implemented using an ELT layout to enhance radiation tolerance to TID effects. M4 is a cascode PMOS that improves the gain and additionally isolates the input from the output node. Because the source-follower (M3) and the transconductor (M1) share the same current, the circuit is power-efficient. The input node can be biased and reset using different schemes that include a diode reset, a PMOS reset or a leakage compensation circuit.

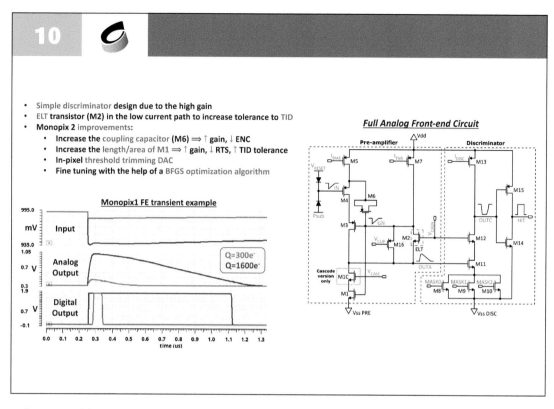

- Simple discriminator **design due to the high gain**
- **ELT transistor (M2) in the low current path** to increase tolerance to TID
- **Monopix 2** improvements:
  - **Increase the** coupling capacitor **(M6)** ⟹ ↑ gain, ↓ ENC
  - **Increase the** length/area of M1 ⟹ ↑ gain, ↓ RTS, ↑ TID tolerance
  - **In-pixel** threshold trimming DAC
  - **Fine tuning of a** BFGS optimization algorithm

The preamplifier is followed by a discriminator which is implemented by a common-source stage that operates as a current comparator. Such a simple discriminator implementation is made possible due to the very high gain of the combined sensor/preamplifier which is higher than 0.5 mV/e-. The Monopix2 front-end transistors have been further optimized for minimum noise and threshold dispersion. The coupling MOS capacitor (M6) has been enlarged by approximately 7.5 times to increase the preamplifier bandwidth and improve gain and noise performance. Furthermore, the length of M1 has been increased to reduce its output conductance and increase the voltage gain. As a result of the higher gain, the threshold dispersion due to the discriminator (as seen from the pre-amplifier input) is significantly reduced. An in-pixel threshold trimming DAC has been also included to further improve threshold uniformity.

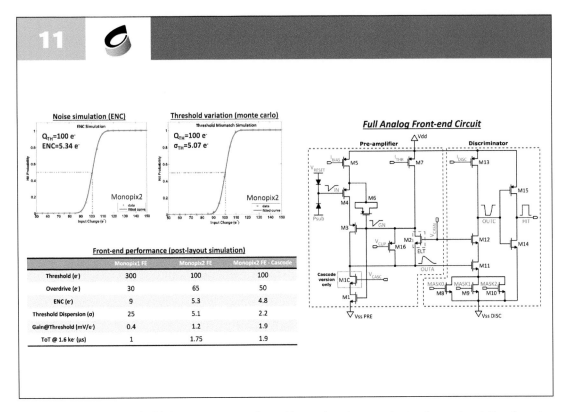

N oise (ENC) and threshold dispersion are evaluated by performing transient noise and variation (monte-carlo) simulations of the front-end response for different input charge values and calculating the hit probability (probability of the discriminator firing) for each charge value. The resulting plot has the shape of a so-called S-curve which is formulated as cumulative distribution function of a normal distribution. RMS noise in electrons and threshold dispersion are calculated by applying a fit to the corresponding S-curve. The ENC of TJ-Monopix2 has been reduced to 5e- (from 9e-) while its threshold dispersion has been reduced by approximately 5 times (from 25e- to 5e-) compared to TJ-Monopix1 without an increase in power consumption.

12

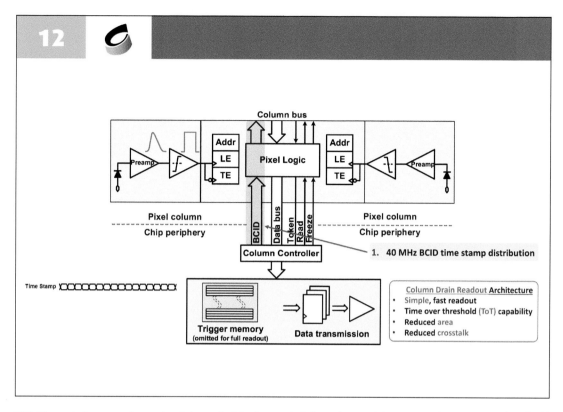

1. **40 MHz BCID time stamp distribution**

Column Drain Readout **Architecture**
* Simple, **fast readout**
* **Time over threshold (ToT) capability**
* **Reduced** area
* **Reduced** crosstalk

TJ-Monopix incorporates a standalone, fast "column-drain" readout architecture with Time-over-Threshold (ToT) information. It has been selected due to its simplicity (low area, reduced crosstalk) and hit rate capability which complies with the ATLAS ITk outer layer specifications (~100 Mhz/cm2). Its operation is as follows: A 7-bit (6-bit in the case of TJ-Monopix1) bunch crossing ID (BCID) time stamp is distributed across the pixel matrix. The BCID is gray encoded in order to reduce crosstalk and power consumption (1 transition/period). When a HIT pulse is produced by the discriminator of a pixel, its leading edge (LE) and trailing edge (TE) time-stamp are recorded in SRAM memories in the pixel. The pixel data includes the LE, TE time-stamp and the pixel address which is encoded in a ROM. Subsequently, a busy token flag is raised and propagates through the column. A dedicated readout controller at the chip periphery receives the token signal and responds with two control signals called READ and FREEZE. FREEZE prohibits new hits from disturbing the current readout cycle and READ initiates the readout operation. All pixels share a common bus, which can be accessed by one pixel at a time with priority from top to bottom. The pixel that has been hit and has the highest priority is granted access to the bus when the READ signal is received. The pixel data is received by the end-of-column block and latched in the digital periphery. A similar priority scheme is implemented across the matrix columns while the column address is appended to the pixel data in the periphery. The readout cycle continues until all pixels that have been hit are read out. Pixel data including the position (address), timing (LE) and charge information (ToT=TE-LE) is directly transmitted off-chip to the data acquisition (DAQ) board.

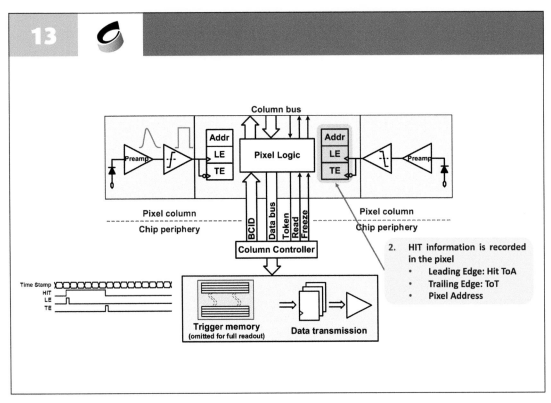

TJ-Monopix incorporates a standalone, fast "column-drain" readout architecture with Time-over-Threshold (ToT) information. It has been selected due to its simplicity (low area, reduced crosstalk) and hit rate capability which complies with the ATLAS ITk outer layer specifications (~100 Mhz/cm2). Its operation is as follows: A 7-bit (6-bit in the case of TJ-Monopix1) bunch crossing ID (BCID) time stamp is distributed across the pixel matrix. The BCID is gray encoded in order to reduce crosstalk and power consumption (1 transition/period). When a HIT pulse is produced by the discriminator of a pixel, its leading edge (LE) and trailing edge (TE) time-stamp are recorded in SRAM memories in the pixel. The pixel data includes the LE, TE time-stamp and the pixel address which is encoded in a ROM. Subsequently, a busy token flag is raised and propagates through the column. A dedicated readout controller at the chip periphery receives the token signal and responds with two control signals called READ and FREEZE. FREEZE prohibits new hits from disturbing the current readout cycle and READ initiates the readout operation. All pixels share a common bus, which can be accessed by one pixel at a time with priority from top to bottom. The pixel that has been hit and has the highest priority is granted access to the bus when the READ signal is received. The pixel data is received by the end-of-column block and latched in the digital periphery. A similar priority scheme is implemented across the matrix columns while the column address is appended to the pixel data in the periphery. The readout cycle continues until all pixels that have been hit are read out. Pixel data including the position (address), timing (LE) and charge information (ToT=TE-LE) is directly transmitted off-chip to the data acquisition (DAQ) board.

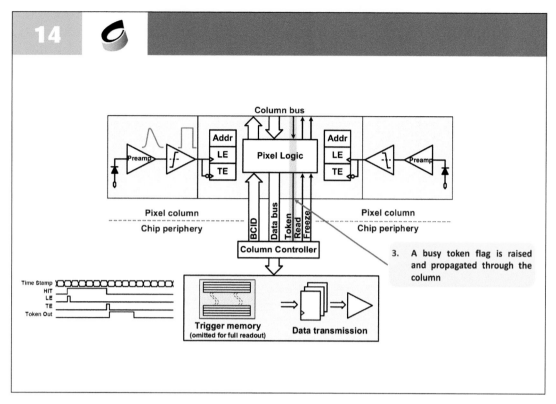

3. A busy token flag is raised and propagated through the column

TJ-Monopix incorporates a standalone, fast "column-drain" readout architecture with Time-over-Threshold (ToT) information. It has been selected due to its simplicity (low area, reduced crosstalk) and hit rate capability which complies with the ATLAS ITk outer layer specifications (~100 Mhz/cm2). Its operation is as follows: A 7-bit (6-bit in the case of TJ-Monopix1) bunch crossing ID (BCID) time stamp is distributed across the pixel matrix. The BCID is gray encoded in order to reduce crosstalk and power consumption (1 transition/period). When a HIT pulse is produced by the discriminator of a pixel, its leading edge (LE) and trailing edge (TE) time-stamp are recorded in SRAM memories in the pixel. The pixel data includes the LE, TE time-stamp and the pixel address which is encoded in a ROM. Subsequently, a busy token flag is raised and propagates through the column. A dedicated readout controller at the chip periphery receives the token signal and responds with two control signals called READ and FREEZE. FREEZE prohibits new hits from disturbing the current readout cycle and READ initiates the readout operation. All pixels share a common bus, which can be accessed by one pixel at a time with priority from top to bottom. The pixel that has been hit and has the highest priority is granted access to the bus when the READ signal is received. The pixel data is received by the end-of-column block and latched in the digital periphery. A similar priority scheme is implemented across the matrix columns while the column address is appended to the pixel data in the periphery. The readout cycle continues until all pixels that have been hit are read out. Pixel data including the position (address), timing (LE) and charge information (ToT=TE-LE) is directly transmitted off-chip to the data acquisition (DAQ) board.

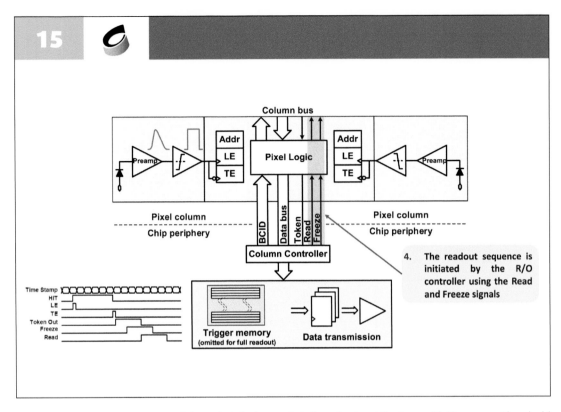

TJ-Monopix incorporates a standalone, fast "column-drain" readout architecture with Time-over-Threshold (ToT) information. It has been selected due to its simplicity (low area, reduced crosstalk) and hit rate capability which complies with the ATLAS ITk outer layer specifications (~100 Mhz/cm2). Its operation is as follows: A 7-bit (6-bit in the case of TJ-Monopix1) bunch crossing ID (BCID) time stamp is distributed across the pixel matrix. The BCID is gray encoded in order to reduce crosstalk and power consumption (1 transition/period). When a HIT pulse is produced by the discriminator of a pixel, its leading edge (LE) and trailing edge (TE) time-stamp are recorded in SRAM memories in the pixel. The pixel data includes the LE, TE time-stamp and the pixel address which is encoded in a ROM. Subsequently, a busy token flag is raised and propagates through the column. A dedicated readout controller at the chip periphery receives the token signal and responds with two control signals called READ and FREEZE. FREEZE prohibits new hits from disturbing the current readout cycle and READ initiates the readout operation. All pixels share a common bus, which can be accessed by one pixel at a time with priority from top to bottom. The pixel that has been hit and has the highest priority is granted access to the bus when the READ signal is received. The pixel data is received by the end-of-column block and latched in the digital periphery. A similar priority scheme is implemented across the matrix columns while the column address is appended to the pixel data in the periphery. The readout cycle continues until all pixels that have been hit are read out. Pixel data including the position (address), timing (LE) and charge information (ToT=TE-LE) is directly transmitted off-chip to the data acquisition (DAQ) board.

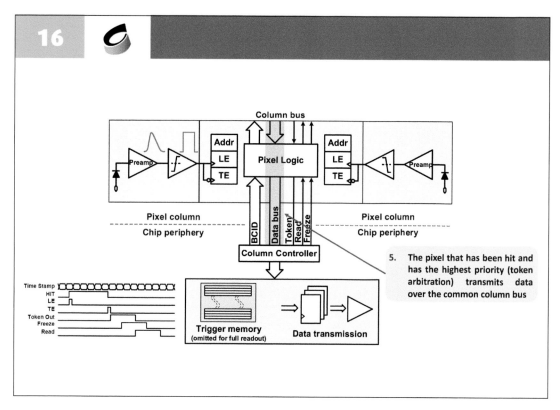

5. The pixel that has been hit and has the highest priority (token arbitration) transmits data over the common column bus

TJ-Monopix incorporates a standalone, fast "column-drain" readout architecture with Time-over-Threshold (ToT) information. It has been selected due to its simplicity (low area, reduced crosstalk) and hit rate capability which complies with the ATLAS ITk outer layer specifications (~100 Mhz/cm2). Its operation is as follows: A 7-bit (6-bit in the case of TJ-Monopix1) bunch crossing ID (BCID) time stamp is distributed across the pixel matrix. The BCID is gray encoded in order to reduce crosstalk and power consumption (1 transition/period). When a HIT pulse is produced by the discriminator of a pixel, its leading edge (LE) and trailing edge (TE) time-stamp are recorded in SRAM memories in the pixel. The pixel data includes the LE, TE time-stamp and the pixel address which is encoded in a ROM. Subsequently, a busy token flag is raised and propagates through the column. A dedicated readout controller at the chip periphery receives the token signal and responds with two control signals called READ and FREEZE. FREEZE prohibits new hits from disturbing the current readout cycle and READ initiates the readout operation. All pixels share a common bus, which can be accessed by one pixel at a time with priority from top to bottom. The pixel that has been hit and has the highest priority is granted access to the bus when the READ signal is received. The pixel data is received by the end-of-column block and latched in the digital periphery. A similar priority scheme is implemented across the matrix columns while the column address is appended to the pixel data in the periphery. The readout cycle continues until all pixels that have been hit are read out. Pixel data including the position (address), timing (LE) and charge information (ToT=TE-LE) is directly transmitted off-chip to the data acquisition (DAQ) board.

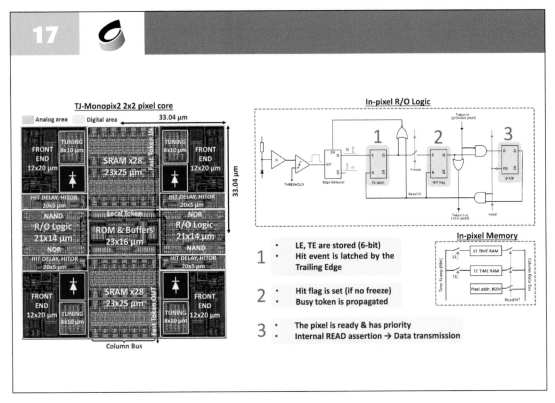

The TJ-Monopix2 pixel size is 33.04x33.04 μm2. The analog part contains the collection electrode, pre-amplifier, discriminator and threshold trimming DAC. The digital part contains 14 SRAM cells/pixel, the in-pixel column-drain readout logic and test features such as a HITOR signal. The discriminator HIT pulse is processed by an edge detector which produces the TE and LE short pulses that are connected to the SRAM write input. The hit event is registered by a first latch activated by the TE pulse and the edge detector is disabled. If the column is not frozen (the FREEZE signal is not asserted), the HIT flag, stored in a second latch, is set and the readout sequence begins. The HIT flag activates the token signal that propagates through the column and informs the R/O controller that data is ready to be read out. At the same time, it prohibits lower priority pixels to access the column data bus during the read phase. At the rising edge of READ, the state of the pixel that has been hit and has the highest priority is stored in a D-latch. If the D-latch is set, and while READ is active, an internal read (READINT) signal is produced that enables access to the data-bus.

18

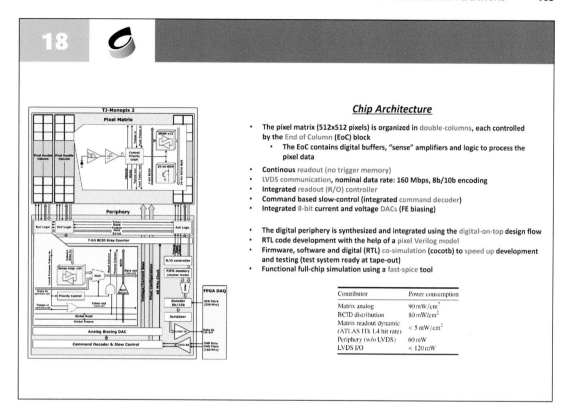

Chip Architecture

- The pixel matrix (512x512 pixels) is organized in double-columns, each controlled by the End of Column (EoC) block
  - The EoC contains digital buffers, "sense" amplifiers and logic to process the pixel data
- Continous readout (no trigger memory)
- LVDS communication, nominal data rate: 160 Mbps, 8b/10b encoding
- Integrated readout (R/O) controller
- Command based slow-control (integrated command decoder)
- Integrated 8-bit current and voltage DACs (FE biasing)

- The digital periphery is synthesized and integrated using the digital-on-top design flow
- RTL code development with the help of a pixel Verilog model
- Firmware, software and digital (RTL) co-simulation (cocotb) to speed up development and testing (test system ready at tape-out)
- Functional full-chip simulation using a fast-spice tool

Contributor	Power consumption
Matrix analog	90 mW/cm²
BCID distribution	80 mW/cm²
Matrix readout dynamic (ATLAS ITk L4 hit rate)	< 5 mW/cm²
Periphery (w/o LVDS)	60 mW
LVDS I/O	< 120 mW

The TJ-Monopix2 pixel matrix is composed of 512x512 pixels and is organized in double-columns. Each double column is controlled by an end-of-column (EoC) block which contains digital buffers to drive the BCID timestamp and control signals, sense amplifiers to readout the data-bus and logic to arbitrate between columns and process the pixel data. Since no trigger memory is implemented, all pixel data is transmitted off-chip directly by an LVDS transmitter at 160 Mbps after being 8b/10b encoded. Chip control is achieved by an integrated command decoder. The periphery also includes 8-bit current and voltage DACs that are used to bias the front-end circuit. The digital periphery is described in RTL (Verilog), synthesized and integrated using the digital-on-top design flow. A pixel Verilog model has been used in order to accelerate the digital periphery RTL code development and verification. The test setup (firmware, software) and digital periphery RTL has been co-simulated using the cocotb framework to speed up the development of the test system and minimize errors.

19

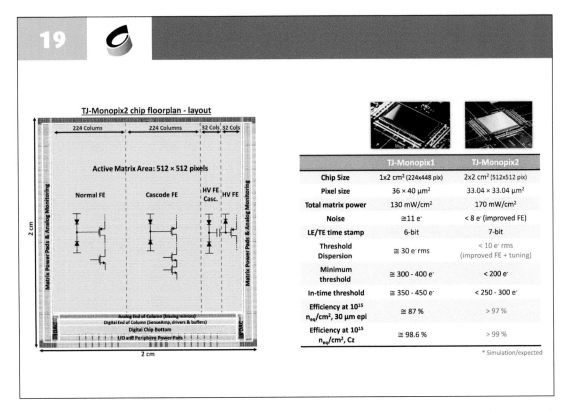

TJ-Monopix2 chip floorplan - layout

	TJ-Monopix1	TJ-Monopix2
Chip Size	1x2 cm² (224x448 pix)	2x2 cm² (512x512 pix)
Pixel size	36 × 40 μm²	33.04 × 33.04 μm²
Total matrix power	130 mW/cm²	170 mW/cm²
Noise	≅11 e⁻	< 8 e⁻ (improved FE)
LE/TE time stamp	6-bit	7-bit
Threshold Dispersion	≅ 30 e⁻ rms	< 10 e⁻ rms (improved FE + tuning)
Minimum threshold	≅ 300 - 400 e⁻	< 200 e⁻
In-time threshold	≅ 350 - 450 e⁻	< 250 - 300 e⁻
Efficiency at 10^{15} n_{eq}/cm^2, 30 μm epi	≅ 87 %	> 97 %
Efficiency at 10^{15} n_{eq}/cm^2, Cz	≅ 98.6 %	> 99 %

* Simulation/expected

The main characteristics of the TJ-Monopix1 and TJ-Monopix2 are presented and can be easily compared. The power consumption of the pixel matrix is low (<200mW/cm2) due to the power-efficient front-end amplifier as a result of the low sensor capacitance. The matrix power includes the analog power consumption (90mW/cm2) and the BCID distribution power consumption (80mW/cm2). The matrix power consumption of TJ-Monopix2 is higher due to its smaller pixel size.

20

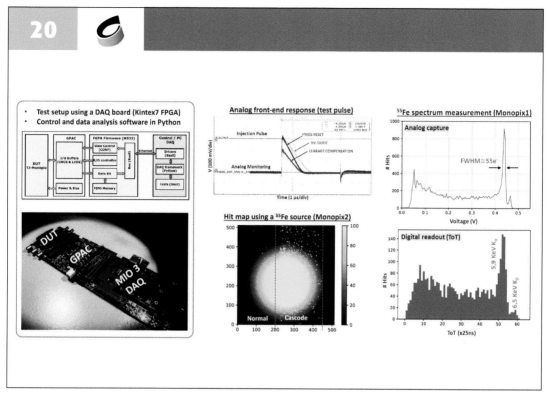

The test setup is based on a DAQ board using a Kintex7 FPGA. Control and data analysis is performed using dedicated software written in Python. The analog front-end response can be measured using special analog-monitoring pixels at the side of the pixel matrix that include a suitable buffer. Individual pixels can be tested using a variable amplitude test pulse that drives an in-pixel charge injection capacitance. In order to calibrate the injection capacitance and additionally assess the front-end performance, TJ-Monopix has been illuminated with a 55FE radioactive source with produces X-ray photons at 5,9 keV (Kα) and 6,5 keV (Kß). The source illumination area can be observed by a HIT map that is recorded using TJ-Monopix2. A 55FE spectrum measurement using TJ-Monopix1 analog monitoring pixels and digital readout (ToT) is also provided. The two peaks (Ka, Kb) are clearly visible, which is indicative of the high signal to noise ratio (SNR) of the TJ-Monopix pixel detector.

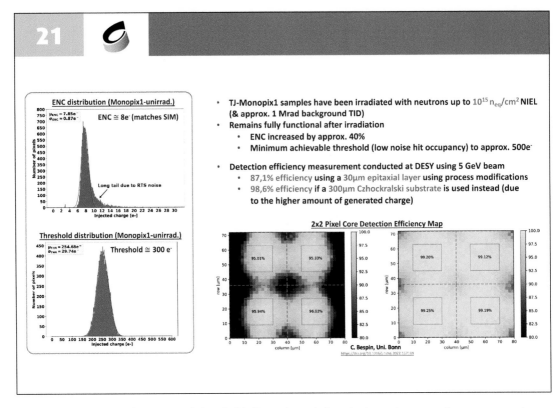

In order to measure the ENC and threshold dispersion of the pixel matrix, several measurements are conducted while varying the test pulse amplitude i.e. injected charge value. The HIT probability is calculated for each input charge value and a fit is performed on the resulting S-curve. The ENC and threshold values that are extracted for each individual pixel are subsequently histogrammed and a gaussian fit is applied in order to calculate the mean and rms values. Here measurements from TJ-Monopix1 are presented because its characterization is complete. The mean ENC is approximately 8e-, which is very close the simulation result. The threshold dispersion is measured to be equal to approx. 30e- while the operating threshold (mean) was set to 255 e-. TJ-Monopix1 samples have been irradiated with neutrons up to a NIEL fluence of 10^15 neq/cm2 in order to assess its radiation tolerance to bulk damage by measuring the detection efficiency. The detection efficiency is measured in test beam facilities using a telescope. The telescope is composed of several reference planes (pixel detectors) before and after the device under test (DUT) and is used to reconstruct the beam particle tracks. The efficiency is simply calculated as the ratio of the particles detected by the DUT (TJ-Monopix) to the total number of particles tracks. A test beam measurement was conducted at the Deutsches Elektronen-Synchrotron (DESY) with a 5 GeV electron beam. The mean detection efficiency of TJ-Monopix1 with process modifications and a 30µm epitaxial layer substrate was measured after irradiation to be approx. 87%. The loss of efficiency mainly occurs at the pixel edges due to the relatively low electric drift field and can be reduced by shrinking the pixel size, lowering the threshold or increasing the input signal. TJ-Monopix2 was designed to address these improvements by reducing the pixel size by 25% and the operating threshold by about 2-3 times. The input signal can be increased by a larger sensor volume which leads to a higher amount of generated charge carriers. If a thicker Czhockralski substrate is used instead of the 30µm epitaxial layer, the detection efficiency of TJ-Monopix1 has been measured to be 98,6%.

PART II: CDR/PLL

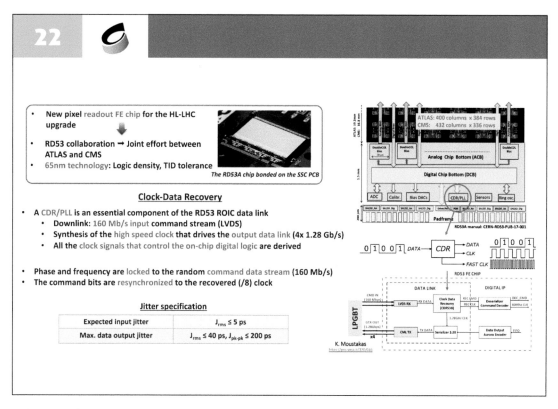

The RD53A chip bonded on the SSC PCB

- **New pixel** readout FE chip **for the HL-LHC upgrade**
- **RD53 collaboration** → Joint effort between **ATLAS and CMS**
- **65nm technology:** Logic density, TID tolerance

Clock-Data Recovery

- A CDR/PLL **is an essential component of the RD53 ROIC data link**
  - **Downlink:** 160 Mb/s input **command stream (LVDS)**
  - **Synthesis of the** high speed clock **that drives the** output data link **(4x 1.28 Gb/s)**
  - **All the** clock signals that control the on-chip digital logic **are derived**

- **Phase and frequency are** locked to the random command data stream **(160 Mb/s)**
- **The command bits are** resynchronized to the recovered (/8) clock

Jitter specification

Expected input jitter	$J_{rms} \leq 5$ ps
Max. data output jitter	$J_{rms} \leq 40$ ps, $J_{pk-pk} \leq 200$ ps

The second part of the presentation is dedicated to communication and data link of a pixel detector and more specifically the Clock-Data Recovery (CDR)/PLL block which is a critical for the ROIC operation as it generates the reference for all digital clock signals on the chip. In this context, a radiation hard CDR/PLL circuit is presented that was designed for the new ATLAS and CMS ROIC (RD53) which will be installed during the high luminosity LHC upgrade. The task of the CDR/PLL is to recover the reference clock from the random 160 Mb/s input command data stream and synthesize the high-speed clock (1.28 Ghz) that drives the output data link. All other clock signals are derived for the high-speed clock by division. The input command data is resynchronized to the recovered clock and is subsequently read by the command decoder. One of the most important specifications of the CDR/PLL is the jitter budget in order to achieve a stable good quality link. In this application the input jitter (downlink) is expected to be approximately around 5 ps rms, while the output jitter (uplink) should be lower than 200 ps peak-to-peak.

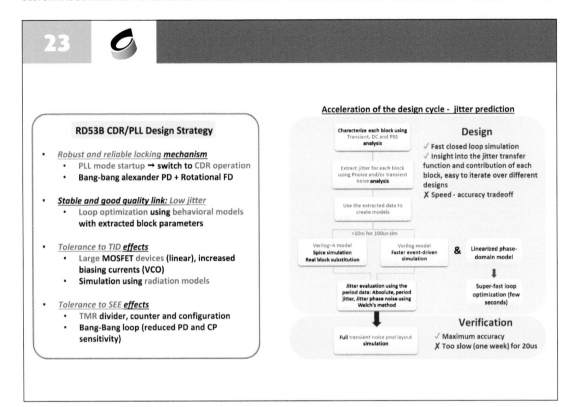

The CDR/PLL should always achieve lock reliably. Because a CDR has a quite narrow lock-in range and can also lock to harmonics or the reference clock, initial locking is achieved in PLL mode by using a standard phase-frequency detector (PFD) while a training clock sequence is provided at the input. After a pre-determined amount of time, the operation is switched to CDR mode and the input command data starts to flow. In CDR mode, a bang-bang alexander phase detector (PD) is used which is aided by a rotational frequency detector. The working principle of the alexander PD is the sampling of the input data with three consecutive edges of the divided (feedback) clock, two positive and one negative. If the first two samples have the value of "0" and the third is "1", the divided clock is early. If on the other hand the last two samples have the value of "1" and the first sample is "0", the divided clock is late. The rotational frequency detector has a pull-in range of approx. 25% of the input frequency and guarantees that locking will be restored if the phase deviation becomes large due to e.g. a single event transient (SET) or excessive noise. In order to accelerate the closed loop simulation and optimize jitter efficiently, behavioral models have been developed in VerilogA. Each block has been simulated and optimized using SPICE and its phase noise/jitter was extracted. The phase noise/jitter parameters were then imported into the model in order to tune the loop parameters for minimum jitter. Following this approach, the full closed loop simulation (transient noise) time has been reduced from a few days to under 10 minutes without significant loss in accuracy. Radiation tolerance to TID effects has been enhanced by using large transistor geometries and increased bias currents in sensitive blocks such as the voltage controlled amplifier (VCO). The CDR operation has been additionally validated using radiation transistor models developed by the collaboration. Finally, susceptibility to single event effects (SEE) has been reduced by employing triple modular redundancy (TMR) in the design of critical blocks such as the frequency divider.

The CDR/PLL forms a second order feedback loop which is characterized by its bandwidth and damping factor. For frequencies below the CDR/PLL bandwidth, the loop is tracking the input and acts as a low pass filter with respect to the input phase noise/jitter power spectral density. Thus, in this region the input jitter is not filtered. In the frequency band above the CDR/PLL bandwidth, the VCO jitter will accumulate (act more like a free-running oscillator) because the loop is not fast enough to track the input phase. Therefore, in this region the CDR/PLL acts as a high pass filter with respect to the VCO phase noise/jitter power spectral density. The CDR/PLL bandwidth is set to an optimal value taking into account the expected input jitter and VCO performance. The loop filter parameters must be tuned for maximum phase margin in order to avoid jitter peaking close to the cutoff frequency. Additionally, because in the case of bang-bang loops the phase detector is "binary" (only provides information about the sign of the phase error), the resulting output frequency step must be considered and minimized.

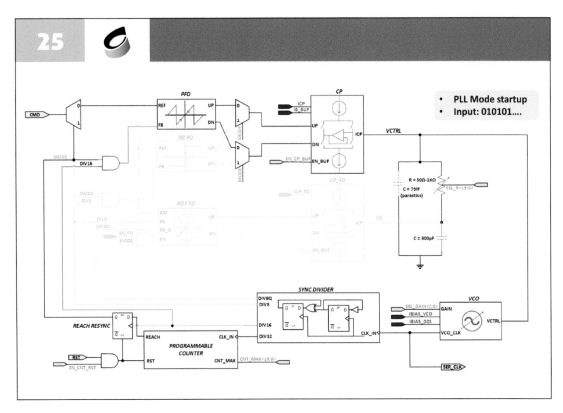

The CDR/PLL loop starts in PLL mode, while a training pattern (80 MHz clock) is provided at the input. In this mode the PFD is active while the alexander PD is disconnected. A programmable counter, clocked using a divided by 32 version of the VCO clock starts to count down. When the counter value becomes 0, CDR mode is activated. The charge pump input is switched to the alexander PD and the rotational frequency detector is also enabled. In normal operating mode, the CDR/PLL loop consumes about 6mW of power. The loop filter resistor value, VCO gain and charge pump current can be digitally controlled in order to tune the CDR/PLL bandwidth.

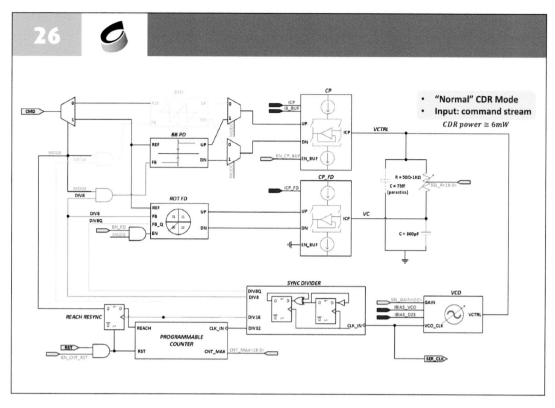

The CDR/PLL loop starts in PLL mode, while a training pattern (80 MHz clock) is provided at the input. In this mode the PFD is active while the alexander PD is disconnected. A programmable counter, clocked using a divided by 32 version of the VCO clock starts to count down. When the counter value becomes 0, CDR mode is activated. The charge pump input is switched to the alexander PD and the rotational frequency detector is also enabled. In normal operating mode, the CDR/PLL loop consumes about 6mW of power. The loop filter resistor value, VCO gain and charge pump current can be digitally controlled in order to tune the CDR/PLL bandwidth.

27

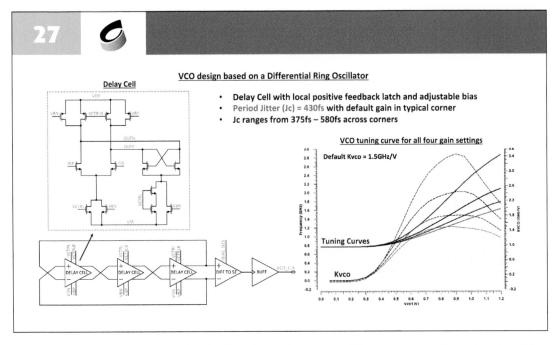

VCO design based on a Differential Ring Oscillator

Delay Cell

- Delay Cell with local positive feedback latch and adjustable bias
- Period Jitter (Jc) = 430fs with default gain in typical corner
- Jc ranges from 375fs – 580fs across corners

VCO tuning curve for all four gain settings

Default Kvco = 1.5GHz/V

Tuning Curves

Kvco

The voltage controlled oscillator (VCO) is based on a three stage differential ring oscillator (DRO) which is followed by a differential to single-ended converter/buffer. The delay cell consists of a differential pair with active load and incorporates a cross-coupled positive feedback latch to improve its performance. The VCO gain is digitally adjustable and its period jitter is equal to 430fs.

28

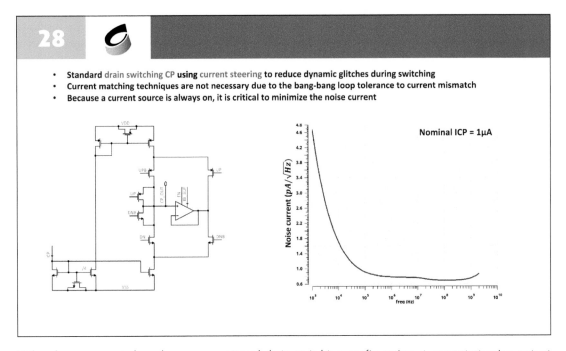

- Standard drain switching CP **using** current steering **to reduce dynamic glitches during switching**
- **Current matching techniques are not necessary due to the bang-bang loop tolerance to current mismatch**
- **Because a current source is always on, it is critical to minimize the noise current**

Nominal ICP = 1µA

The charge-pump is based on a conventional drain-switching configuration. A current steering output stage using a unity-gain buffer is employed to suppress spurious signals due to charge sharing during switching. Advanced current matching techniques are not necessary because the bang-bang loop is tolerant to current mismatch. However, since in a bang-bang loop the charge-pump is always sourcing or sinking current, it is critical to minimize its noise current.

29

- Synchronous **divider with outputs: /16 (PFD), /8 (PD), /8 Quad (FD), /32 (CDR/PLL mode counter)**
- **Only the last stage contributes to jitter: simulated edge-to-edge jitter:** J_{ee} = 117.5 fs
- **Fully triplicated logic (DFF & combinational gates for SEE robustness)**
- **Modified 12 track LVT library to increase TID tolerance**

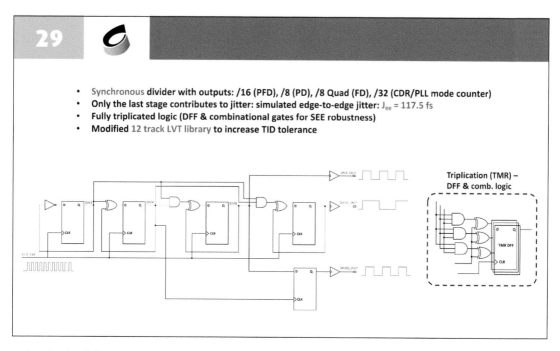

The divider follows a synchronous configuration in order to reduce its jitter contribution (jitter is only determined by the last stage). A modified 12 track digital library which uses large transistor geometries has been used in the divider design to improve tolerance to TID radiation effects. Each stage is fully triplicated (TMR) followed by a voter gate to enhance robustness to single event effects (SEE).

30

- A dedicated **mini-ASIC** test chip (2x2 mm²) was built to characterize the RD53B CDR performance that includes:
    - LVDS command receiver
    - Serializer, PRBS generator & CML GTX fast data link cable driver
    - Bandgap, biasing DAC's
    - Triple modular redundant (TMR) SPI configuration

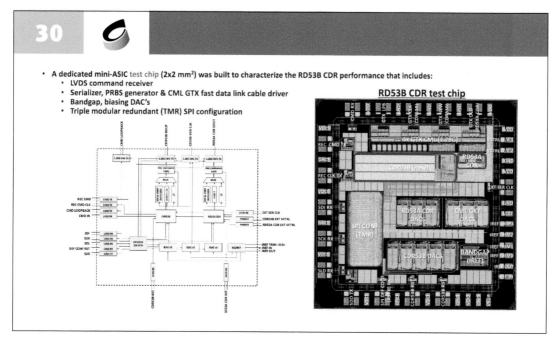

A mini-ASIC test chip has been developed as a test vehicle for the CDR/PLL block. It contains all the necessary supporting blocks such as biasing DACS and SPI slow control. The input clock/data stream is provided by an LVDS receiver. The CDR/PLL output can be directly monitored and is also used to drive a serializer which can be loaded with a configurable pseudo-random binary sequence (PRBS). A gigabit CML driver with configurable pre-emphasis is used to drive the output signal.

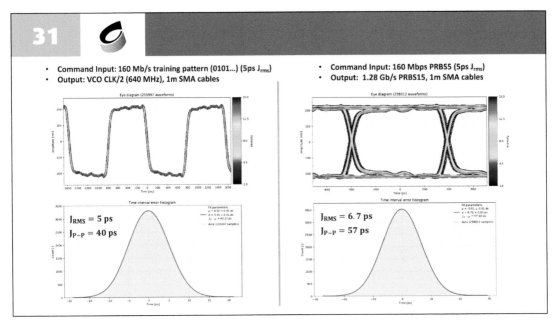

The CDR/PLL jitter performance was measured with the help of a setup wired with 1m long SMA cables. A 160 Mb/s data stream with 5ps rms jitter was provided at the input and the chip was configured to output serialized PRBS15 data at 1.28 Gb/s, which approximates the real data output stream in the experiment. In this configuration, the measured CDR output jitter for a representative large sample of 250k edges was measured equal to 6.7 ps rms and 57 ps peak-to-peak, which is well withing the jitter budget specification.

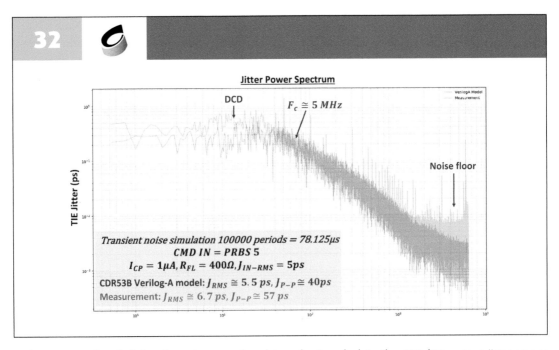

The measured period data have been analyzed in order to calculate the CDR/PLL output jitter power spectrum and validate the behavioral model simulation. Very good matching between simulation and measurement can be observed. The deviation at high frequencies is caused by the oscilloscope noise floor, while the difference at low frequencies is attributed to duty cycle distortion of the input signal, which generates a dead zone due to the alexander PD operation.

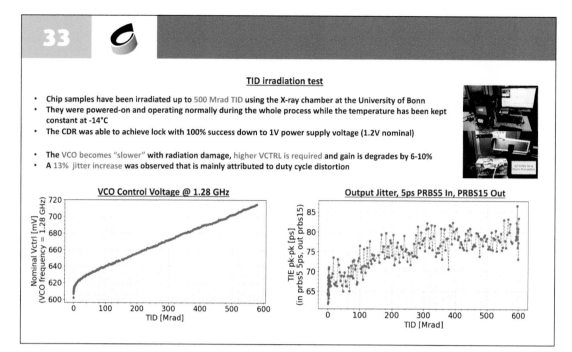

The CDR/PLL test chip was irradiated with X-rays up to a total TID of 600 Mrad to assess its radiation tolerance. The circuit remained fully operational and was able to achieve lock with 100% success. The VCO control voltage increases with radiation and indicates that the VCO becomes "slower" mainly due to the TID influence on the delay cell PMOS load transistors. The output jitter has increased by only 13%, however this increase is attributed to duty cycle distortion in the test chip output signal path.

CHAPTER 08

Why is a Verification Engineer the Most Wanted Person these Days?

Olivera Stojanovic

VTOOL, Ltd., Serbia

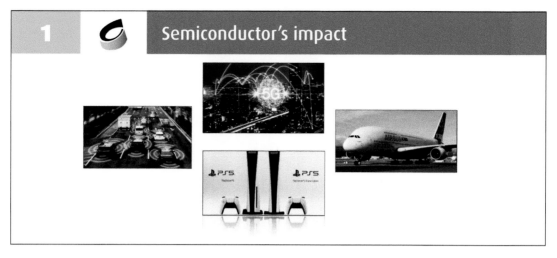

L et's see what are new technologies
L .   ·   As first example, one of the hottest topics are autonomous cars. Beside the main idea that they are driver less, plan is that they will communicate between each other. In this area, Tesla is going even further; they claim that in the future, while you are at home, your car can be used as a driverless taxi. In order words, your car will make money while you are resting at home.
· 5G network will make possible new applications in *internet-of-things* (IoT) and *machine-to-machine* areas. As an example sensor in yours fridge in the kitchen, detects that you do not have milk any more, automatically sends request to shop and ship it to your home. Moreover, all these steps are executed without human interaction.
· Avionics industry I can not put into new technology, but there are improvements there. One of the examples would be controlling via GPS path of airplane while he is on runway, in order to have most optimal path and decrease time spend on runway.
· And the last one. If you ask my oldest son what is newest technology for him. He will tell you for sure that it is Playstation 5.
Beside these examples, there are other areas where new technologies can have impact on our daily lives: medicine, internet of Behavior, face recognition, Artificial inelegant, Virtual Reality and its implementation in different areas: industrial, medicine, learning, playing.
All of these are new technologies, and all of them requires different types of chips to support them. On this slide I presented some of the projects that I was involved with, working on development chips.

L et's see a few facts about semiconductor industry.
L

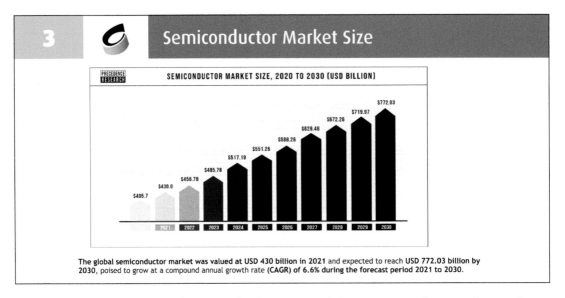

The global semiconductor market was valued at **USD 430 billion in 2021** and expected to reach **USD 772.03 billion by 2030**, poised to grow at a compound annual growth rate **(CAGR) of 6.6%** during the forecast period 2021 to 2030.

The global semiconductor market was valued at USD 430 billion in 2021 and expected to reach USD 772.03 billion by 2030. It is one of the most profitable industries in the world.

Increase in demand on chips last several years lead to short supply globally for nearly two years. You probably do not know, but the biggest semiconductor manufacturer is TSCM and it is based on Taiwan.

We all remember this year tensions related to Nancy Pelosi. Why did she decide to visit Taiwan at such a sensitive time in China-U.S. relations? Was there a point to this trip? Speaker of the House's controversial visit, she made sure to meet with one of the self-governing island's most important business leaders, Mark Liu, chairman of Taiwan Semiconductor Manufacturing Co. (TSMC), the world's biggest chipmaker. The discussions came just days after Congress passed the Chips and Science Act, which provides $52 billion in subsidies to incentivize chip manufacturers to build factories in the United States. TSMC is expected to receive a chunk of those subsidies to help fund manufacturing facilities it is building in Arizona.

So, we can also see that beside US investing $52 billion for semiconductor EU launches $43b push for chip factories as shortages hit manufacturing.

We can all see that biggest players in the world see the importance of chip industry and that they are investing heavily in it.

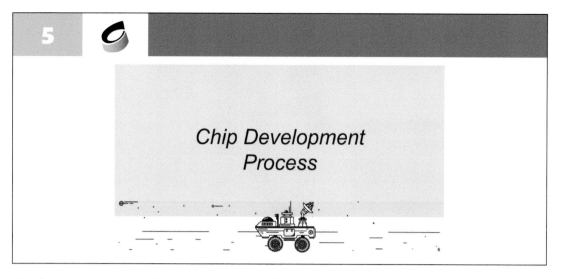

So, let see how chip development process looks like.

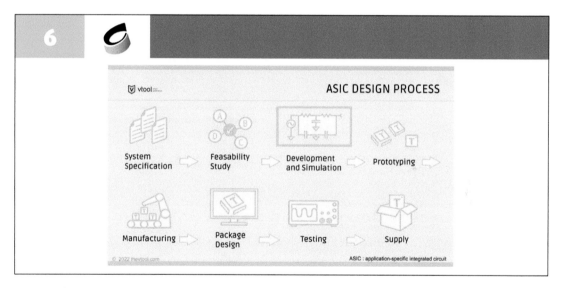

1. .As a first step is to define, what will be the functionality of the chip. Main person in this stage is architect of the chip whose main responsibility is to define architecture of the chip.
2. After that feasibility study is performed. During this process, we need to check if this chip will satisfy targeted performances with proposed architecture. Also, since development of the chip is very expensive, we need to calculate targeted volume according to need on the market and decide if this would be valuable business.
3. As a next step, we would proceed with development and simulation. During this process, team of engineers is working on it. Main goal of this stage is to develop and verify chip before it goes to production. Bugs found in later stages can introduce significant costs due to respin or even louse market share if we are too late compare to competitors.
4. After this stage we are developing prototype and test it in lab. If there is some bug found, we may return to previous stage.
5. When we have final chip we are moving to fabrication and manufacturing.
6. Each chip is than put to the package.
7. When we have physical chip, additional testing is performed on the testers. Main purpose of this stage is to check if there is any defect during fabrication or during packaging.
8. And the last stage is shipping to the customer.

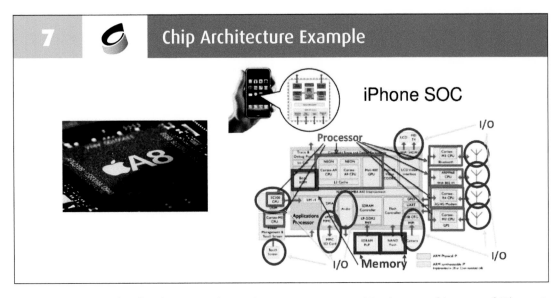

This is one example of architecture of one chip. You can see on this picture architecture of iPhone 8. We can see some main components like processors, interconnect, peripherals, memories... In order to develop this chip, it requires different profiles of engineers. : digital designers, verification engineers, backend and analog engineers. Beside them in some cases, we have algo and sw engineers.

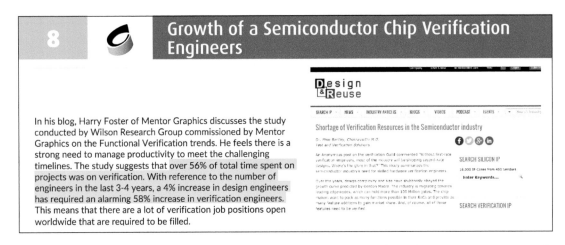

Over the years, we see significant increase in design complexity . The industry is migrating towards leading edge nodes, which can hold more than 100 Million gates. The chip makers want to pack as many functions possible in their SoCs and provide as many feature additions as possible to gain market share. And, of course, all of those features need to be verified.

A Gartner study suggests that the key industry trends in the Global Chip Design market are: Escalating cost of chip design, increasing design complexity and shorter time for new product launch. The worst thing that could happen to a Chipset company is a respin due to a buggy product. This calls for greater concerted effort to verify the complex designs and a particular need of skilled and smart verification resources to help thwart increasing development costs.

Verification engineers requires strong analytical skills and creative thinking. There are numerous scenarios to think through not only at system/chip level but also at the application level, where things could really go wrong. With all this effort the verification team is helping their company save millions of dollars. Hence there is a greater challenge in making the design work than just designing itself.

Due to all these facts, the study conducted by Wilson Research Group commissioned by Mentor Graphics on the Functional Verification trends shows that over 56% of total time spent on projects was on verification and required an alarming 58% increase in verification engineers in next 3-4 years.

L ets check the main concepts of constraint random verification.

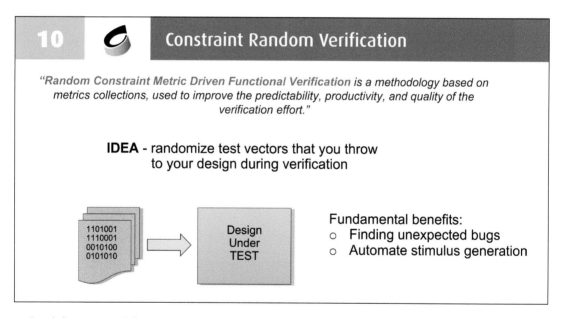

The definition say follow: "Random Constraint Metric Driven Functional Verification is a methodology based on metrics collections, used to improve the predictability, productivity, and quality of the verification effort."

The main IDEA in constraint random verification using random vectors in simulation. Over the last few years random verification represent state of the art how to use simulation. Idea is to randomize test vectors that you throw to your design.

There are two main benefits in using random verification:

1. First fundamental benefit that random verification is great for finding unexpecting bugs. Just think, when you are planning to have all combinations of inputs, then you are going to find only bugs that you are looking for. Idea with constrain random verification is to randomly generate traffic and put Design Under test in state that you did not think of.

2. Other key benefit is automatic simulation generation. Point about automotive simulation is that you can run long simulation (during night, weekends) without writing manually test vectors by yourself, they will be generated automatically.

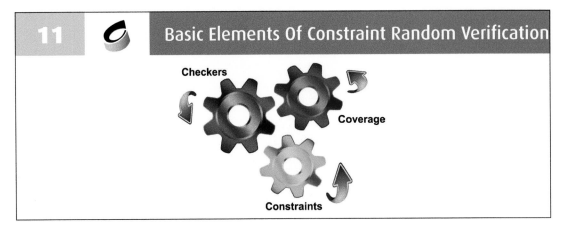

| **11** | | **Basic Elements Of Constraint Random Verification** |

There are 3 main elements of constraint random verification, we used to call them 3 C's:
CHECKERS: If you have random vectors, it is critical to have automatic self-checking testbenches. Checkers will give us answer to question: Does design works?
Two types of checking:
1. Temporal checking - assertions   2. Data Checking – scoreboard checking
COVERAGE: If you have random vectors, you must have metric that can show you how many of design features have been testes. Coverage will give us answer to the questions: Are we done yet?
Two types of coverage:
1. Code Coverage   2. Functional Coverage
CONSTRAINTS: will help us to generate appropriate input to the design and help us to get random stimulus that is matching some conditions according to specification. They are also valuable for hitting corner cases in the design as well for generating error scenarios.

Simulator itself, contains constraint solver and mechanism to generate different stimulus according to the seed and therefore we will have automated generation of various scenarios.

| **12** | | **Languages and Methodology** |

❖ SystemVerilog – UVM methodology
❖ Specman e- eRM methodology

The most popular languages and methodologies in verification are System Verilog with UVM methodology and Specman-e with eRM Methodology.
Verification Methodology represents:
· Standardized framework for coverage-driven functional verification
· Promotes reuse   · Self-checking testbench
· Transaction-level modeling   · Coverage metrics...
· By following methodology is very easy to understand code and find something, even if you are not developer of that code.
Verification languages features:
· OOP approach   · Constrained random generation
· Modular, scalable and reusable   · Assertion based verification
  generic verification environment   · Coverage driven verification

Let's see how to build verification environment.

14 — VIP - Verification Intellectual Property

VIP: Verification Intellectual Property
- Reusable
- Pre-verified
- Configurable
- Plug and Play

Main building blocks of each verification environment are VIPs.
VIP means Verification Intellectual Property and main key characteristics are:
- Reusability (if you can reuse something that for sure save you time ,and if you safe time you are saving money in this business).
- Second characteristic is Pre-Verified. When you are developing VIP there are some steps that you can do in order to test it before you use it in module level verification. After that you have high level of confidence that VIP is working correctly.
- Next one is characteristic is configurability – VIP is developing to be configurable, that means that every feature according to Protocol Specification that can have different values depending on certain implementation – user specific VIP must support. For example, different with of data or number of lanes.
- And the last one is plug and play – when you have developed VIP, it is very easy to connect it to the DUT and instantiate it in verification environment. You can even use predefined test, sequences or even error injection scenarios that are already part of VIP package

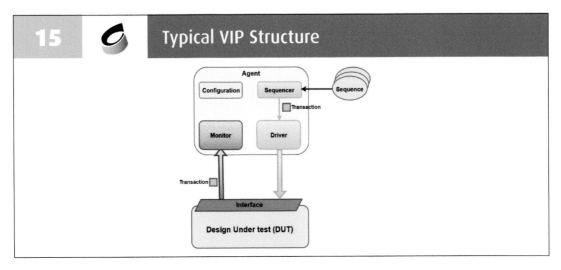

Methodology gives us guidelines how to build VIP and this is typical structure of one VIP. The main part of VIP is agent. Inside of agent thre are:

- Config
- Monitor
- Sequencer
- Driver
- Virtual Interface

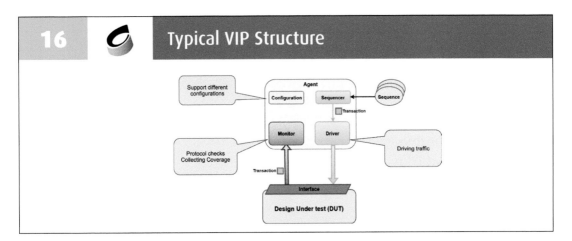

Transaction is abstract representation of one transfer on interface. It is class that contains relevant information of one transfer.

Virtual interface is module where all relevant signals for this interface are defined and it is used for connecting to appropriate ports of the DUT.

Monitor is totally passive part of VIP. When I say passiv, I mean that there is no interaction between DUT and VIP. It is just monitoroing lines on interface , collect and check.

Monitor functions:

- Collecting transactions – Monitor is collecting transaction according to protocol and have appropiate hooks that can be used for scoreboards on module level verification
- The protocol checks: comparing the interface signals' behavior to the behavior specified by the protocol
- The coverage collection - used as a tracker of the interface usage completeness in means of interface capabilities

Driver is active part of the VIP and it is used for getting transaction from sequencer and drive it to DUT

By UVM methodology there is mechanism how driver, sequencer and sequence operate. Basically, in the test we are using sequences, sequences are generating transactions and at the end driver will be responsible to translate this object to pin driving.

Config- supports different configurations.

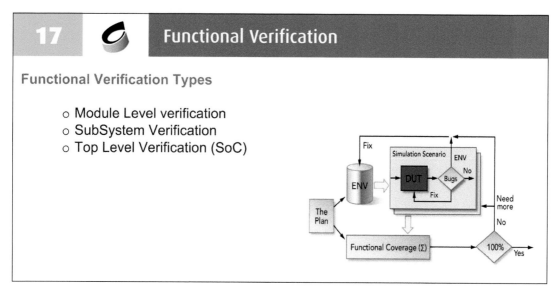

Functional Verification Types

- o Module Level verification
- o SubSystem Verification
- o Top Level Verification (SoC)

Functional verification can be done on several levels: module level, subsystem verification and top level verification.

Random constrain metric driven verification promotes reuse on two levels: horizontal and vertical. Reuse of common components from project to project or reuse components from lower to higher level.

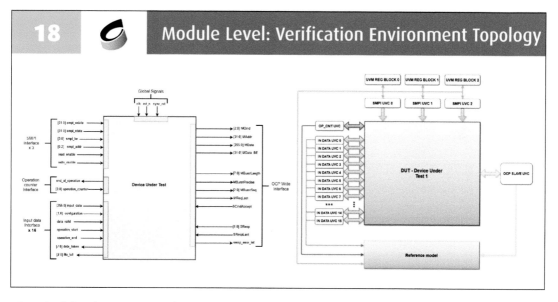

On the left side you can see the example of one module that I tuck from one of my projects that I was working on (real life example).

So how you shall start building a verification enviroment? When you are doing module level verification , first thing you do is to define all interfaces that that this Moule have. Usually you will have for each of interface one instance of dedicated VIP. VIP can be developed for that module or reuse. In this example SMPI and OCP UVCs are reused from UVC library that we already had from previous modules and two others are developed.

. VIP will be used for driving stimulus, creating tests, checking if DUT is aling with protocol, monitoring and collect transaction . Transaction will be pass than to prediction who will do prediction according to functionality that DUT shall do. Predicted transaction than will be forwarded to Scoreboard. On other side, other VIP will collect transactions , give response if necessary to DUT and send transaction in Scoreboard. Scoreboard will then check if predicted transaction is same as received one.

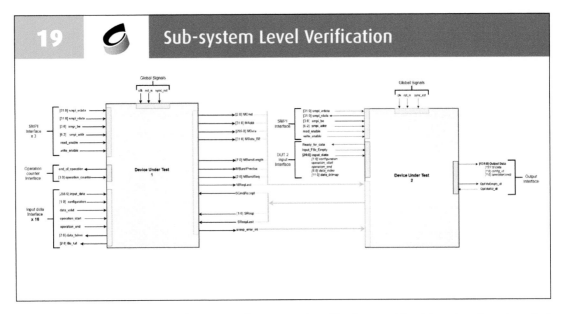

N ew level of verification is sub-system. The scope of verification on this picture is sub-system that contains two modules.

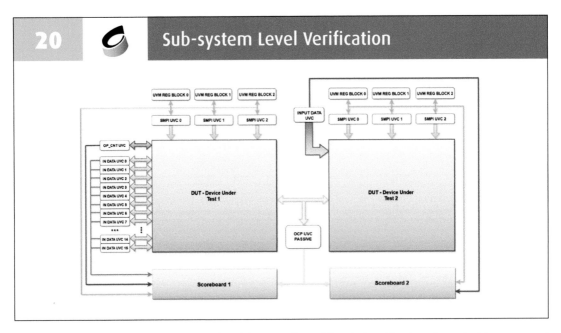

W hen moving to a higher subsystem level (eventually top) level of verification, the only thing removed from a VIP usage is the signal driving for inter-connections of the modules.

Protocol checks, coverage, and scoreboards (for data path checks) remains the same at higher levels of verification.

If we tested each of this DUTs as separete modules according to methodology it is very easy to reuse both enviroments . Only thing that you need to do is to is change connections and put VIPs that are now in middle to be passive. You can even reuse some of the sequences or tests that you use on module level verification.

If the module level verification is done properly, the top level verification phase should be straight-forward and rather simple

21   SoC Level Verification

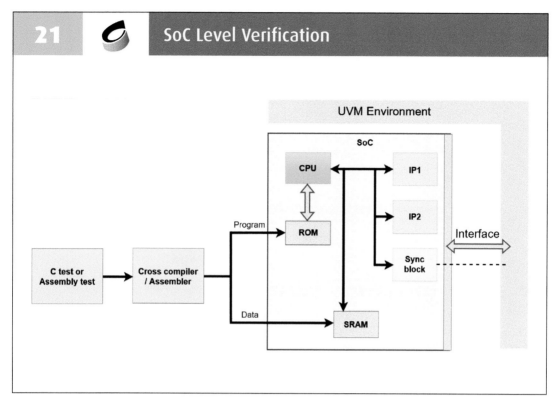

Each SoC, by definition, contains one or more processors and the code that they execute represents a significant part of the overall functionality. We need also to verified together hardware and software in order to check full range of operations.

While on module level focus was on black box testing of functionality and reaching each line and condition in design code, on SoC level is focus on verifying all connections between the blocks are functioning correctly and that data flows properly around the SoC. Also one of the features that we want to test on SoC is that all hardware interrupts and associate interrupt service routine in software are working correctly.

To perform SoC verification we need several things:
· Build UVM verification environment
· Create C code that will be executed on CPU
· Converting C programs to binary files for the embedded processor(s)
· Synchronizing C programs and UVM tests

The testbench can also help the firmware and software engineers write and debug device drivers and applications.

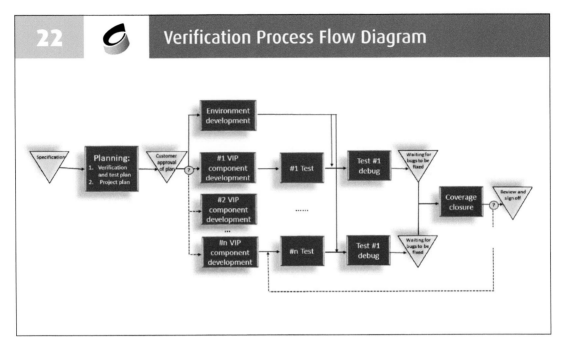

1) **Specification:** customer provides specification where chip functionality is defined.
2) **Project plan:** Verification engineer creates full project and verification plan according to the specification

3) **Customer approval of the plan:** customer reviews the project plan and in few iterations with verification engineer make final version

4) **Environment development:** verification environment development can start at the beginning of the project, but it cannot be finalized before all VIP components are developed

5) **VIP components development:** each chip has set of dedicated interfaces. Each interface is modeled in the appropriate VIP component.

6) **Tests development:** each test cannot be developed before the dedicated VIPs are ready

7) **Test debug:** when test is implemented, debug process starts. In most of the cases, bugs are found during chip debug process, and these tasks will end when all of the bugs are fixed and tests pass

8) **Coverage closure:** when all tests pass, coverage closure task starts and it includes analyzing matrix that represents portion of the code that has been tested so far. For each line in the design that has not been covered additional tests are written. This process is considered to be done when 100% coverage is achieved and all added tests are passing.

9) **Review and sign off:** When regression results are ok (all tests passing and 100% coverage reached) customer performs review and officially signs off.

23 Chip Verification

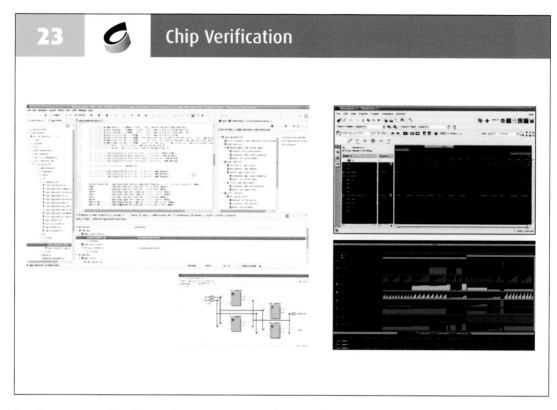

Let's see practically what verification engineer is doing and recap some things that we said so far.

As verification engineer your main responsibility is to test and stress the chip, also try to think about scenarios that designer did not think about it, but also to be sure that designer understood specification correctly. In practice, what you will need to do as verification engineer:

1. Understand functional specification and create verification plan that includes verification environment topology, checkers, coverage and test definition

2. We are developing of verification environment using programming languages like SystemVerilog or Specman. During this process, we need to follow methodology that is called UVM, that promotes best methods of random verification, reusability, modularity, standards that everyone understands. In essence, we need to develop code using object-oriented languages.

3. The main goal is to create stimulus on inputs of the chip, and check if design is performing correct functionality on the outputs.

4. During this process we need to use different tools. For writing code, we are using editors. When code is ready, we are using simulators to check on the waveform behavior of the chip as well if our stimulus is correct.

5. When the test failed, we need to determinate where is the issue- eider bug in design or in our verification environment. For that purpose, we are extensively using messages that are printed in the log files.

6. In verification community it is very well known that debug process is very slow, hard, requires high engineering skills and the most problem part is that you can not predict how long it will take. We, at the Vtool recognize this pain and decided to develop tool that is call Cogita. Cogita is the debugging platform that is helping verification engineer to grasp only relevant information, easy manipulate with the log and by visual presentation helps verification engineer to find much faster root cause of the failing test.

7. Sign off criteria, or when we say that verification is done, is when we have all tests passing, no open bugs, 100% code and functional coverage.

24 — Career Advice: How to choose the company you would like to work for?

Define your personal values:

- I want to learn new things constantly, make progress and share my knowledge
- I want to be part of something that will make an impact in the work I do
- I want to be surrounded by people who share the same passion for work

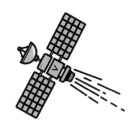

Choose your professional career and company that shares your personal values!

25 — Vtool

Business units:

1. **Services**
   - Digital design
   - Verification
   - Embedded
2. **EDA development – Cogita**

Locations:

1. Israel
2. Serbia
3. Greece
4. Poland
5. France

www.thevtool.com

About
the Editors

Thomas Noulis

Thomas Noulis is an Academic and a Semiconductor Industry expert (INTEL Corp., INFINEON AG, HELIC Inc. (acquired by ANSYS)) with more than 12 years of international experience on Design System Development, Analog/RFMS circuit Design, Design Methodology, and international project management. He holds B.Sc. Degree in Physics, MSc. Degree in Electronics Engineering, and a Ph.D. from Aristotle Univ. of Thessaloniki in collaboration with LAAS, Toulouse-France.

He has participated in multiple Mobile communications and Power automotive product development Integrated Circuit Design projects in Academia and in the Industry Sector and he has extensive Tape out experience in advanced semiconductor technologies from FinFET and RFCMOS to CMOS, Power Automotive BCD and Bipolar processes. Thomas Noulis is the main author of more than 60 publications in international journal and conferences, he holds one European patent, and he is the Editor of the books "Mixed Signal Circuits" and "Noise Coupling in System on Chip" by CRC Press. He is an active reviewer and a member of the editorial board of international scientific journals and conferences. He has given an invited presentation on System on Chip crosstalk and Readout from end IC design. His research interests are focused on Design System Development, Design Cycle Speed Up, Low noise circuit design, Signal and Power Integrity, Instrumentation, and sensor interfaces. Currently he is an Assistant Professor in the Physics Dept. of Aristotle Univ. of Thessaloniki, Greece. He is a Senior Member of IEEE.

Costas Psychalinos

Costas Psychalinos received his B.Sc. and Ph.D. degrees in Physics and Electronics from the University of Patras, Greece, in 1986 and 1991, respectively. From 1993 to 1995, he worked as Post-Doctoral Researcher in the VLSI Design Laboratory, University of Patras. From 1996 to 2000, he was an Adjunct Lecturer at the Department of Computer Engineering and Informatics, University of Patras. From 2000 to 2004 he served as Assistant Professor at the Electronics Laboratory, Department of Physics, Aristotle University of Thessaloniki, Greece. From 2004 he served as faculty member at the Electronics Laboratory, Department of Physics, University of Patras, Greece and, currently, he is full Professor.

His research area is in the development of CMOS analog integrated circuits, including fractional-order circuits and systems, continuous and discrete-time analog filters, amplifiers, and low voltage/low power building blocks for analog signal processing.

He serves as Editor-in-Chief of the Circuit and Signal Processing Section of the MDPI Electronics Journal. He serves as Area Editor of the International Journal of Electronics and Communications (AEUE) Journal, and Editor of the International Journal of Circuit Theory and Applications. He is Associate Editor of the Circuits Systems and Signal Processing Journal, and the Journal of Advanced Research. He is member of the Editorial Board of the Microelectronics Journal, Analog Integrated Circuits and Signal Processing Journal, Fractal & Fractional Journal, and Journal of Low Power Electronics and Applications.

He is IEEE Senior Member and, also, Member of the Nonlinear Circuits and Systems Technical Committee of the IEEE CAS Society.

Alkis A. Hatzopoulo

Alkis A. Hatzopoulos (Chatzopoulos) received his Degree in Physics (with honours), his Master Degree in Electronics and his Ph.D. Degree in Electrical Engineering from the Aristotle University of Thessaloniki, Greece, in 1980, 1983 and 1989, respectively. He has been with the Department of Electrical and Computer Engineering at the Aristotle University of Thessaloniki since 1981, were now he is a full Professor. Since 2002 he has been elected as the Director of the Electronics Laboratory of the ECE Dept.

His research interests include modelling, design and testing of integrated circuits and systems (analog, mixed-signal, high-frequency), three dimensional Integrated Circuits (3D ICs), electronic communication circuits, thin-film transistors, instrumentation electronics, Bioelectronics systems, space electronics.

He is actively involved in educational and research projects, and he is the author or co-author of more than 170 scientific papers in international journals and conference proceedings and three book chapters in international textbooks. He has been also granted a European and American patent.

He has been a Member (2010-2018) of the Belgian Research Foundation Committee (FWO Expert panel WT7) for National (Flanders) Grants and Project evaluations. He has given a large number of lectures on his research topics in various Universities. He has been a visiting Professor at Michigan State University, MI, USA, (Feb. - July 1995), at the Katholieke Universiteit Leuven, Belgium (Feb. - July 2004), and at the University of Texas at Dallas (UTD) (Jan. - July 2016). He is a Senior Member of IEEE and he has served as the IEEE Greece CASS-SSCS joint Chapter Chair from 2010 to 2019.

About
the Authors

CHAPTER 1

Shahram Minaei received the B.Sc. degree in Electrical and Electronics Engineering from Iran University of Science and Technology, Tehran, Iran, in 1993 and the M.Sc. and Ph.D. degrees in electronics and communication engineering from Istanbul Technical University, Istanbul, Turkey, in 1997 and 2001, respectively. He is currently a Professor at the Department of Electronics and Communications Engineering, Dogus University, Istanbul, Turkey. He has more than 180 publications in scientific journals or conference proceedings. His current field of research concerns current-mode circuits and analog signal processing. Dr. Minaei is an associate editor of the Journal of Circuits, Systems and Computers (JCSC), and editor-in-chief of the AEU-International Journal of Electronics and Communications.

CHAPTER 2

Costas Psychalinos received his B.Sc. and Ph.D. degrees in Physics and Electronics from the University of Patras, Greece, in 1986 and 1991, respectively. From 1993 to 1995, he worked as Post-Doctoral Researcher in the VLSI Design Laboratory, University of Patras. From 1996 to 2000, he was an Adjunct Lecturer at the Department of Computer Engineering and Informatics, University of Patras. From 2000 to 2004 he served as Assistant Professor at the Electronics Laboratory, Department of Physics, Aristotle University of Thessaloniki, Greece. From 2004 he served as faculty member at the Electronics Laboratory, Department of Physics, University of Patras, Greece and, currently, he is full Professor.

His research area is in the development of CMOS analog integrated circuits, including fractional-order circuits and systems, continuous and discrete-time analog filters, amplifiers, and low voltage/low power building blocks for analog signal processing.

He serves as Editor-in-Chief of the Circuit and Signal Processing Section of the MDPI Electronics Journal. He serves as Area Editor of the International Journal of Electronics and Communications (AEUE) Journal, and Editor of the International Journal of Circuit Theory and Applications. He is Associate Editor of the Circuits Systems and Signal Processing Journal, and the Journal of Advanced Research. He is member of the Editorial Board of the Microelectronics Journal, Analog Integrated Circuits and Signal Processing Journal, Fractal & Fractional Journal, and Journal of Low Power Electronics and Applications.

He is IEEE Senior Member and, also, Member of the Nonlinear Circuits and Systems Technical Committee of the IEEE CAS Society.

CHAPTER 3

Paul. P. Sotiriadis, Fellow IEEE, is a Professor of Electrical and Computer Engineering of the National Technical University of Athens, Greece, the Director of the Electronics Laboratory of the University and a governing board member of the Hellenic (National) Space Center of Greece.

He runs a team of 30 researchers. He received the Diploma degree in Electrical and Computer Engineering from same University, the M.S. degree in Electrical Engineering from Stanford University, USA and the Ph.D. degree in Electrical Engineering and Computer Science from the Massachusetts Institute of Technology, USA, in 2002. In 2002, he joined the faculty of the Johns Hopkins University Electrical and Computer Engineering Department and in 2012 he joined the faculty of the Electrical and Computer Engineering Department of the National Technical University of Athens. He has authored and coauthored more than 200 research publications, most of them in IEEE journals and conferences, holds one patent, and has contributed several chapters to technical books.

Prof. Sotiriadis research interests include the design, optimization, and mathematical modeling of analog, mixed-signal and RF integrated and discrete circuits, sensors and instrumentation architectures with emphasis in biomedical instrumentation, advanced RF frequency synthesis, and, the application of machine learning and general AI in the operation as well as the design of electronic circuits.

He has received several awards, including the prestigious Guillemin-Cauer Award from the IEEE Circuits and Systems Society in 2012, a Best Paper Award in the IEEE International Conf. on Microelectronics (ICM), 2021, Best Paper Award in the IEEE Symposium on Integr. Circ. and Sys. Design (SBCCI), 2021, Best Paper Award in the IEEE International Conf. on Microelectronics (ICM), 2020, Best Paper Award in the IEEE International Conf. on Modern Circ. and Sys. Tech. 2019, Best Paper Award in the IEEE International Frequency Control Symposium 2012, Best Paper Award in the IEEE International Symp. on Circuits and Systems 2007 and the IEEE Circuits and Systems Society (CASS) Outstanding Technical Committee Recognition 2022.

Dr. Sotiriadis is an Associate Editor of the IEEE Sensors Journal, has served as an Associate Editor of the IEEE Trans. on Circuits and Systems – I (2016-2020) and the IEEE Trans. on Circuits and Systems – II (2005-2010) and has been a member of technical committees of many conferences. He regularly reviews for many IEEE transactions and conferences and serves on proposal review panels.

CHAPTER 4

Georgios Zervakis received the Diploma and Ph.D. degrees from the School of Electrical and Computer Engineering (ECE), National Technical University of Athens (NTUA), Greece, in 2012 and 2018, respectively. He is currently an Assistant Professor at the Department of Computer Engineering & Informatics, University of Patras, Greece. Before that he was a Research Group Leader at the Chair for Embedded Systems (CES), at the Karlsruhe Institute of Technology (KIT) from 2019 to 2022. From 2015 to 2019, Dr. Zervakis worked in many EU-funded research projects as a research associate of the Institute of Communication and Computer Systems (ICCS), Athens, Greece. Dr. Zervakis serves as a reviewer in many IEEE and ACM Transactions journals, including TCAS-I, TVLSI, TCAD, TC, and is also a member of the technical program committee of several major design conferences, including DAC, DATE, and ASP-DAC. Dr. Zervakis has published more than 40 papers (in peer-reviewed journals and conferences), including 19 IEEE Transactions publications, and 4 book chapters in Springer. He has received one best paper nomination at DATE 2022. Dr. Zervakis is a leading expert in approximate computing for circuit design and specifically for machine learning circuits. His main research interests include low-power design, accelerator microarchitectures, approximate computing, design automation, printed electronics and machine learning.

CHAPTER 5

Hesham Omran received the B.Sc. (with honors) and M.Sc. degrees from Ain Shams University, Cairo, Egypt, in 2007 and 2010, respectively, and the Ph.D. degree from King Abdullah University of Science and Technology (KAUST), Saudi Arabia, in 2015, all in Electrical Engineering. From 2008 to 2011, he was a Design Engineer with Si-Ware Systems (SWS), Cairo, Egypt, where he worked on the circuit and system design of the first miniaturized FT-IR MEMS spectrometer (NeoSpectra), and a Research and Teaching Assistant with the Integrated Circuits Lab (ICL), Ain Shams University. From 2011 to 2016 he was a Researcher with the Sensors Lab, KAUST. He held internships with Bosch Research and Technology Center, CA, USA, and with Mentor Graphics, Cairo, Egypt. In 2016, he rejoined the ICL, Ain Shams University, where he is currently an Associate Professor. He created the Mastering Microelectronics YouTube channel with 10k+ subscribers. He co-founded Master Micro in 2020 to develop the Analog Designer's Toolbox (ADT), a novel EDA tool that defines a new paradigm for analog IC design. Dr. Hesham has received several awards including the Egyptian State Encouragement Award for Engineering in 2019, Ain Shams University Encouragement Award for Technology in 2022, and the Design Automation Conference (DAC) Innovator's Award in 2022. He has published 50+ papers in international journals and conferences. His research interests are in the design of analog and mixed-signal integrated circuits, and especially in analog and mixed-signal CAD tools and design automation.

CHAPTER 6

Pedro Filipe Leite Correia De Toledo is an Electronic Engineer with more than ten years of experience developing CMOS RF/Analog Integrated Circuits. Currently, he is a Mixed-Signal IC designer in Synopsys, working in SERDES designs with sub-10nm CMOS technology. He has received Ph.D. at Polito and UFRGS (Joint-Degree Ph.D.) working with Digital-based Analog Processing for IoT Applications. He has finished his M.S. degree in Microelectronics from UFRGS, working with Low Thermal Sensitivity Analog Applications based on ZTC and GZTC MOSFET bias point modeling. He has also received a BSEE (2010), and an RF IC Designer Specialization Course (IC-Brazil Program - 2011) at the Federal University of Pernambuco (UFPE) and the Federal University of Rio Grande do Sul (UFRGS), respectively. He has also won the best 2015 master's thesis in the design, CAD, and IC Test category from the Brazilian Society of Microelectronics (SBMICRO). In 2019, He was awarded the Best Student Paper Award of the 2019 IEEE International Conference on Electronics, Circuits, and Systems (ICECS 2019). Possess a sound understanding of fundamental principles and trends in electronic engineering.

CHAPTER 7

Konstantinos Moustakas is an IC design expert with an interdisciplinary background focused on the enablement of novel pixel sensor technologies, radiation-hard implementations and high-speed serial links. He holds a B.Sc. diploma in Physics, two M.Sc. degrees in Electronics Engineering and Automation Systems and a Ph.D. issued by the University of Bonn.

Dr. Moustakas has a record of successfully developed silicon prototypes in collaboration with leading experts in international projects and was awarded a Marie Sklodowska-Curie fellowship in the framework of the STREAM innovative training network working in the development of smart pixel sensors in collaboration with the CERN microelectronics group. He has published more than 40 articles in peer-reviewed journals and conferences, co-authored a book chapter in the book "Noise Coupling in System on Chip" and is an active reviewer in the Journal of Electronics and Communications (AEUE).

His research interests include pixel detector IC's, low noise readout front-ends, low voltage/low power techniques, radiation tolerant designs and Phase Locked Loop (PLL)/Clock Data Recovery (CDR) circuits. In 2021 he has joined the Photon Science Division (PSD) Detector Group at the Paul Scherrer Institut (PSI, Villigen AG, Switzerland) and his work is currently focused on the new generation of pixel detectors for X-ray applications.

CHAPTER 8

Olivera Stojanović holds MSc in electrical engineering from the School of Electrical Engineering in Belgrade and EMBA from Cotrugli Business School.

Her diverse experience is concentrated on project management and managing teams of verification engineers. She spent a portion of her career helping senior management and the sales department through business development management. Her expertise in the technological field has helped her to better comprehend and anticipate customer needs.

Being the product owner of the Cogita tool, which is largely concerned with debugging and assisting verification engineers in their endeavors, as well as managing the Embedded business unit, is Olivera's greatest accomplishment and current area of competence. She highlights the value of its features and devotes her varied knowledge to its upgrading as her primary goal in developing and establishing the Cogita tool's purpose.

Her main motivation is based on her ability to get the best out of everything she learned throughout her whole career - business development management, verification, creating sales strategies, product design, project management, and leading the group.

Olivera's next goal is to achieve progress in all areas of Cogita development and realize her vision for it to become a valuable tool for fighting bugs.